# Bad UFOs

## Critical Thinking about
## UFO Claims

## Robert Sheaffer

January 10, 2016

**Cover Photo:** The author  created this UFO photo by putting black dots on a cottage cheese container, and placing it on an aluminum plate.

# AUTHOR'S INTRODUCTION

Why write a skeptical book about UFOs?

There has not been a single skeptical book examining the broad UFO phenomenon published in the United States, so far as I know, since my last UFO book *UFO Sightings* (Prometheus Books, 1998). David Clarke's excellent skeptical book, *How UFOs Conquered the World*, was published in the U.K. In 2015.

This book is titled *Bad UFOs*, which is also the name of my Blog *www.BadUFOs.com*. This book hopes to bring the reader up-to-date concerning the most important claims from "the land of UFOria" since the publication of my last book (as Philip J. Klass, the Dean of UFO skeptics, used to call it). This book contains mostly new cases, as well as updates on important older cases.

*The Dean of UFO Skeptics, Philip J. Klass (1919-2005. Photo by author)*

While many people, especially those with scientific training, are skeptical about UFOs, very few actually read skeptical UFO books. And far fewer still bother to *write* a skeptical book on UFOs! Even skeptics who have done first-rate investigations of major UFO cases generally don't bother to write a book describing their work. Klass wrote six such books, one of them for young readers. The last was in 1997. He once explained to me why my first UFO book *The UFO Verdict* would never sell well. "People who are skeptical about UFOs don't need to read your book, they already know that UFOs are nonsense. And the people who believe that UFOs are interplanetary spaceships also don't need to read your book, because they already know

v

you're wrong." I wish I could say Klass was wrong about this, but he wasn't. Fortunately the above doesn't tell the whole story. There are those skeptics who wish to understand the arguments for and against UFOs to better be able to argue their position, simply to be a better skeptic. There are those UFO proponents who want to understand the skeptic's arguments, to better defend *their* position. There are also those who are confused by so many conflicting statements concerning UFOs, and want to read both sides to better make up their own minds. These people are to be commended.

Here everything is linked together into a (hopefully) coherent and up-to-date narrative. All attributions and statements of fact are carefully researched, and can be substantiated. In citing printed works I have provided print references in the References section at the end of this book, since once these are printed, the ink stays in place. However I have decided to handle on-line references in an unusual way. Since most sources discussed in this book can easily be found on-line, we will rely on powerful existing search engines such as Google, Bing, etc. to provide the references for us. Because URLs change frequently, and the links on my *debunker.com* website seem to become outdated almost as soon as I revise them, I have decided not to try to provide exact URLs for most of the references cited in this book. Instead, I have placed some appropriate search terms in boldface type. The reader curious to learn more about the matters discussed here is encouraged to use his or her favorite search engine, with the recommended search terms where given. For example, when I discuss the claims of a UFO crash in Long Island by "**John Ford** of the **Long Island UFO Network** (LIUFON)," put the boldface words into Google and you will have the full background of that strange incident, and its bizarre outcome. An internet search on any person, book, or major UFO case mentioned here will return a great deal of information, not all of it reliable. It will also in many cases return photos or illustrations that I did not include for reasons of copyright. Consider the source in judging the credibility of any UFO claims you encounter, on-line or elsewhere.

Robert Sheaffer, San Diego, California.

# CONTENTS

# ACKNOWLEDGMENTS

No one person can realistically gather together, verify , and present as much UFO information as you will find in this book. There were many who helped me, often providing hard-to-find information when it was most needed.

I'd like to acknowledge Tim Printy, editor of the *SunLite* Webzine, for his prodigious efforts in putting out that magazine, which is a source of copious up-to-date and reliable information about UFO claims. Space analyst and writer James Oberg has been exposing bogus UFO claims for forty years, principally those that involve the space program. James McGaha has a lot of excellent facts, unfortunately he hardly writes anything. We can't forget Ted Molczan, Kitty Mervine, Dave Thomas, and Marty Kottmeyer. On the other side of the Pond, I owe much to Ian Ridpath, Peter Brookesmith, and David Clarke. You will see these people cited frequently in these pages, because they provide excellent quality UFO information.

I'd also like to thank those who, while I might not always agree with them, nonetheless provide important ideas and information that is always worthy of serious consideration: Marc D'Antonio, Jack Brewer, Curt Collins, Shepherd Johnson, Paul Kimball, Isaac Koi, Carol Rainey, Kevin Randle, Chris Rutkowski, Lee Speigel, and Frank Warren.. And I also thank every member of the Roswell Slides Research Group. Thanks, everybody!

Both skeptics and "skeptical believers" agree that the UFO field, as it now stands, is filled to the brim with rubbish. The latter group expects that, when the rubbish is cleared away, there will be a signal in the noise, while the former expects that nothing will be left. But both are natural allies in clearing away UFOlogical rubbish. It also allows us to identify those UFO researchers who are hopelessly mired in delusion, and (for example) still insist that the "Roswell Slides" (chapter 4) do not show a mummy, even after deciphering the placard proclaiming that they do! The real fault line in UFOlogy lies not between skeptics and proponents, but between "UFO realists" – skeptics and skeptical proponents who are willing to look for weaknesses and prosaic explanations for UFO claims, and the "UFO

unrealists" who are ready to accept practically any exciting UFO claim on very little evidence.

# 1. UFOS – SEVEN DECADES, AND COUNTING

What explains the human fascination with UFOs? We must indeed use that term, because interest in the UFO phenomenon has become truly global. Nowadays UFO reports come from every continent, and from almost every major country. The British Fortean skeptic Hilary Evans (1929-2011), who studied the complex relationships between UFO beliefs and social issues, wrote, "It is safe to say that no anomalous phenomenon has generated so rich an anomaly-cluster as the flying saucer" (Evans and Stacy 1997, p. 257). Looking back to the early days of UFO sightings, Evans suggests, "we can see that they were unquestionably an idea whose time had come. We sense an air of inevitability." Indeed. The Flying Saucers came just a few decades after Buck Rogers and Flash Gordon, and just a few years after the V2, Hiroshima, and Nagasaki. We were ready for them, so they must be ready for us. And the "anomaly cluster" Evans mentioned originating from UFO reports now includes the Men In Black, UFO crashes, UFO bases, military and intelligence agency conspiracies, NASA conspiracies, alien abductions, crop circles, alien autopsies, alien-human hybrids, cattle mutilations, and the list just continues to grow.

The first reported sighting of what was then called "flying saucers" was by private pilot Kenneth Arnold on June 24, 1947. Within a few weeks, an entire "wave" of saucer sightings swept across the U.S., and soon across the world. Such "waves" were repeated at intervals of every few years until the last major one in 1973.

Waves of nationwide mass UFO sightings may be a thing of the past, but public interest in UFOs is stronger than ever. Mystery-mongering cable TV programs, the internet, and sensationalized radio talk shows have largely replaced books, magazines, and news media in spreading UFO excitement and misinformation to millions. Belief in UFOs and visitors from other worlds remains high today despite decades of sensational claims unaccompanied by proof. Dubious-looking UFO videos are posted anonymously on *YouTube*, go viral, and for a short time become a topic of conversation across the world, until somebody analyzes it carefully

enough to spot evidence of fakery (and even then, many will not accept the proof that it is a fake).

A child born at the dawn of the UFO era is now a senior citizen. The face of UFOlogy has changed much in these past seven decades, but it has not faded away as some rationalists naively assumed it would. The UFO movement keeps changing, and today the outlook has become largely conspiratorial. Indeed, its mutability is indicative of the strength of the myth, not of its weakness. Let's look at how the UFO movement has changed over the years and what it has become today.

## "New Age" vs. "Science Fiction" UFOlogy

The major fault line among UFO proponents today is the division between what can be called "New Age" UFOlogy and what its proponents call "scientific" UFOlogy but is in reality "science fiction." Both are junk science and consistently ignore Occam's Razor (all other things being equal, the simplest solution is the best). Proponents fail to reconcile whatever hypotheses they invent with the rest of the body of established scientific fact. While the dividing line between the two groups is not hard and fast, and some UFO claims will contain elements of both, most major UFOlogists and UFO groups will fit clearly into one group or the other. "New Age" UFOlogy is dominated by women and "Science Fiction" UFOlogy by men, although you will find members of both genders in either group. We can think of members of the first group as fans of Oprah, the second as fans of the SyFi Channel. "New Age" UFOlogists often seem oblivious to the very idea that anyone should ever have to prove any claims they make, as if people are expected to simply accept unsupported accounts of interactions with unknown beings (as is routinely done in such circles). If you expect to see any kind of proof, you need to hang out in different UFO circles. However, in spite of these deep-seated differences, there is little friction between the two groups, as they seem to have evolved a concept of "non-overlapping magisteria" where each largely ignores the other.

"New Age" UFOlogy grew out of the "contactee" tradition of the 1950s, which is not based primarily on claimed "evidence" but instead on unprovable personal revelations. Contactees reportedly talk to extraterrestrials and receive cosmic wisdom from them, but never are able to offer convincing proof of such communications. Today's "New

Age" UFOlogists largely claim to receive extraterrestrial messages via telepathy, channeling, dreams, or other subjective experiences, continuing the contactee tradition of having a *personal relationship* with the UFOs and their occupants. "New Age" UFOlogy often uses religious terms and themes, typically promoting the idea of an immanent cosmic, metaphysical change in the Earth and in peoples' lives: the "age of Aquarius," the "end of the Mayan Calendar," or some other ill-defined term that largely parallels the concept of the millennium in conventional Christian eschatology.

One well-known group falling squarely in the "New Age" UFO tradition is the *Unarius Educational Foundation* in El Cajon, California. Founded in 1954, the group's members believe that vaguely defined "energies" permeate the universe and claim they receive messages channeled from beings on other planets. They teach that "a new golden age for humanity" will begin as soon as we accept the wisdom and love of our space brethren.

"Science Fiction" UFOlogy claims the reality of visitations from extraterrestrials, or perhaps beings from "another dimension" or some other nebulous realm, based upon the weight of UFO sightings, photos and videos, alleged "trace cases," abductions, UFO crashes, etc. They eagerly offer "proof" when questioned, but it falls short by orders of magnitudes of the evidence required to support such extraordinary claims. They also typically fail to see how their claims contradict accepted science in very significant ways. When they do acknowledge the conflict, they insist it is time to invent a "new" science based upon the "evidence" of UFO incidents, not realizing the impropriety of overturning weighty, well-supported, time-tested scientific principles by anecdotes, as if hummingbird feathers could outweigh elephants.

# UFO Organizations

At the present time, the Mutual UFO Network (MUFON) is the largest and best-known organization of its kind in the U.S. It is primarily made up of "Science Fiction" UFOlogists. It might be most accurate to describe MUFON as "the largest remaining UFO group in the U.S.", since there used to be others of at least its size, which have all folded. Founded in Illinois in 1969 by Walt Andrus (1920-2015) and others, it was originally known as the Midwest UFO Network. Geographically, it was positioned

between its better-known rivals NICAP, headquartered in Washington, DC, and APRO, in Tucson, Arizona. However, each of these UFO groups maintained its own far-flung roster of investigators and "scientific consultants," so that any group might have a presence more or less anywhere. Andrus had originally been affiliated with APRO, but got into a feud with its directors, Coral (1925-1988) and Jim Lorenzen (1922-1986), and struck off on his own. With the demise of its rivals, MUFON found itself the Last Man Standing. It reformulated itself as the Mutual UFO Network, and picked up many of the fading groups' most active and valuable members.

Walt Andrus remained at the helm of MUFON until his retirement in 2000. I met Andrus at the *National UFO Conference* in Phoenix in 1984. He was an irascible man who seems never to have been troubled by any doubts about UFOs, and who was barely able to tolerate skepticism in any form. He described my 1981 skeptical book *The UFO Verdict* as "an insult to the intelligence" of the reader (which delighted me so much that I often used it in the book's promotion).

John Schuessler took over MUFON until his own retirement in 2006, succeeded by a much younger man, James Carrion. I heard Carrion speak to Mensa in Denver in 2008, and chatted with him afterward. Clearly more cautious than Andrus and not so hostile to skeptical questions, Carrion admitted to a great deal of uncertainty concerning UFOs, and would not even make a defense of the Roswell crash claims. His position is that he is sure that UFOs represent something unknown and significant, but does not claim to know what.

Then in late 2009, Carrion resigned from MUFON quite abruptly, as he explained in his Blog posting, "**Goodbye Ufology, Hello Truth.**" His chief complaint is, "What I discovered was that the [UFO] phenomenon is based in deception – of the human kind – and that there is no way ANYONE will understand the real truth unless they are willing to first accept that. No, I am not talking about some grandiose cover-up of alien visitation, but instead the documented manipulation of people and information for purposes that I can only speculate on."

In resigning, Carrion did not hesitate to name names concerning this deception:

2006 – The report by a bogus Dive Company of finding the 1953 Kinross UFO – perpetrators unknown and still at large.

2007 – Michael Nelson's bogus claims of recovering physical evidence related to the 1966 Portage County UFO Chase. Read Nelson's paper in the 2007 MUFON Symposium Proceedings, then tear it out as none of it is based in fact. [A full account of this famous UFO police chase, depicted in the movie *Close Encounters of the Third Kind*, is in my book *UFO Sightings* (Prometheus, 1998)].

2007 – The California Drones story [photos of spiky UFOs] for which not a single, verifiable beyond reproach, real witness can be found.

2008 – The Stan Romanek claims [ET visitation] which are not only unverified by science (despite what he says) but involves the shameful practice of investigators ignoring professional standards by fraternizing and becoming emotionally involved with the subject of their investigation. [Later developments concerning Romanek fully justify Carrion's reservations.]

2009 – Unsettling information I discovered with Dr. Frank Salisbury about the Skinwalker Ranch [a ranch supposedly haunted by paranormal beings] that calls into question the validity of experiences described in the book *Hunt for the Skinwalker*.

I queried Carrion for more information about these incidents, especially the last one, but he did not reply. I realized that Carrion had been skating on thin ice at MUFON, so the fact of his departure was not surprising, although I never expected such a saucy send-off. Carrion concluded, "That in a nutshell is the sad state of Ufology today, humans deceiving humans. If there is a real phenomenon, I have yet to see any evidence of it that would stand under scientific scrutiny."

But organizations like MUFON cannot survive on a fare of UFOlogical caution and caveats. Its members want validation of their passionate belief that extraterrestrials or other exotic beings are visiting us. If they don't get what they want, they'll leave. In the late 1980s and 1990s, MUFON publicly booted out a number of its most prominent investigators for the sin of being too skeptical about one UFO case or another that

# Bad UFOs

Andrus was determined to defend, most notably Ed Walters' absurdly unconvincing hoax UFO photos from Gulf Breeze, Florida. Presumably Andrus found that the publicity over the Gulf Breeze photos was so helpful to MUFON in gaining members that criticism of the case, no matter how solid and factual, was unwelcome within MUFON. But many members of groups like MUFON want "red meat," stories of amazing encounters, abductions, and government conspiracies. Succeeding Carrion as MUFON's International Director was Clifford Clift, a veteran UFOlogist and MUFON field investigator. Clift says that he has been a believer in UFOs since his own sighting of a light in the sky as a child in the early 1950s. "I believe the UFOs exist, they are manufactured objects and they are not of this Earth," he said in a lecture. Clift was clearly better suited to get along much better with MUFON's officers and members.

However, even he encountered difficulties, and many longtime MUFON members criticized what they felt was his autocratic style of management. Clift resigned in 2012, to be replaced by businessman David MacDonald. This involved moving MUFON's entire office and operations from Colorado to Cincinnati, Ohio. One of MacDonald's ventures is the "Mile High Club" operating out of Cincinnati's Municipal Lunken airport. For $425, a couple is promised champagne, chocolates, and privacy as they recline on fluffy cushions while their plane circles the Cincinnati area.

Since August 1, 2013, MUFON's director has been Jan Harzan, and its offices and files are now in Newport Beach, California. "Hangar 1" refers to a building owned by MacDonald in Cincinnati, where some of MUFON's files had been stored. Those files have now been moved to California, but the preposterous MUFON-sponsored cable TV show *Hangar 1* (on the History Channel) ignores that fact, and pretends that the files are so voluminous that an airplane hangar is required to house them. (Since MUFON boasts that it has 70,000 separate files in its computerized case management system, it is not clear why they would need an aircraft hangar to store them.) On *Hangar 1*, MUFON has pulled some of the most discredited stories in all UFOlogy out of the dumpster and hyped them with breathless excitement:

- *President Eisenhower met with aliens!*

- *The Nazis established a flying saucer base in Antarctica!*

- *The U.S. military has captured live aliens from saucer crashes!*

Like Andrus' earlier embrace of the absurd Gulf Breeze photos, many MUFON members and investigators find Hangar 1 absurd and even counter-productive, since it discredits MUFON's claim to be devoted to the scientific investigation of UFO reports. But like Andrus before him with Gulf Breeze, Harzan is obviously finding the popular cable TV program to be invaluable in promoting MUFON, even at the cost of acting just like a crackpot UFO group.

## What Is the Value of UFO Study?

Perhaps you've heard of Bigelow Aerospace, founded by the Las Vegas real estate billionaire Robert Bigelow who made his money with his chain of Budget Suites hotels. But unlike some space entrepreneurs whose plans never leave earth, Bigelow Aerospace has already succeeded in orbiting two of its prototype modules on Russian rockets, Genesis I in 2006 and Genesis II in 2007. These are inflatable modules with sophisticated cameras and electronic packages to demonstrate the feasibility of this unique and untried approach. In 2006, Bigelow Aerospace was awarded the Innovator Award by the Arthur C. Clarke Foundation. In 2013, **Bigelow Aerospace** signed a $17.8 million **contract** with **NASA** to develop an inflatable module for the International Space Station that "folds like a shirt and inflates like a balloon."

But there is one space-related issue troubling Mr. Bigelow, and on which he feels the need to obtain, even at potentially great cost, the best counsel available: UFOs. It is not clear whether he fears that UFOs will interfere with his future orbiting hotel chain, or if he believes that UFOs harbor some secrets of propulsion or anti-gravity that his engineers might someday put to good use. Whichever it is, Bigelow has contracted with MUFON potentially very large sums of money for the pursuit of first-hand UFO information.

Bigelow has a long history in the matter of UFOs and "paranormal" subjects. He was the principal sponsor of the Las Vegas-based National Institute for Discovery Sciences (NIDS) from its founding in 1995 until it was placed on "inactive status" in 2004, and its website apparently has not been updated since then. It reports on a number of UFO investigations, alleged cattle mutilations, and other far-out stuff. The

best-known and most controversial project undertaken by NIDS was its purchase of the supposed **haunted Skinwalker ranch** in northeast Utah, which some describe as a "Hyperdimensional Portal Area" or "Stargate." The ranch is said to be infested by an alien or paranormal shape-shifting creature known as **"Skinwalker,"** taking its name from Native American legends similar to European legends about werewolves. NIDS researchers investigated the ranch starting in 1996. They compiled an impressive collection of what might be termed "ghost stories," but in spite of having access to sophisticated electronic equipment, failed to obtain any proof that anything extraordinary was going on.

*Dr. J. Allen Hynek in his observatory at Northwestern University (photo by author).*

An agreement between Bigelow Aerospace Advanced Space Studies (BAASS) and MUFON sets up a "Star Team Impact Project" (SIP). Investigations will be limited to cases where physical effects of a UFO are reported, or where "living beings" are allegedly sighted, or where "reality transformation" is said to occur. "Lights seen in the sky" do not qualify for paid investigation, a decision with which the Air Force Project Bluebook's scientific consultant Dr. J. Allen Hynek (1910-1986) would have agreed.

In any case, Bigelow is unlikely to turn up anything that is useful for any purpose. Martin S. Kottmeyer wrote a 2013 article **Conflicting Drives** in the *SunLite* E-zine (5-1) detailing the many mutually-contradictory conclusions about 'new physics' one would draw from studying various well-known UFO cases. He concludes,

ufolore is essentially incoherent taken en masse and displays a broad range of ideas. They can't all be right. Not to horrify the reader with hopelessness, Don Keyhoe claimed in 1970 he had read over 200 different scientific and technical explanations of saucer propulsion and he declined to give an opinion on who was right though he felt anti-gravity had interesting explanatory aspects. The larger point is most have to be wrong. Yet, by what independent investigative criterion can anybody prove any of the ideas here are more worth looking into than any of the others? Those who reported propellers and jet engines have credentials as good as anybody and these worthless propulsion concepts showed up more often than the fancier-sounding alternatives. The few fragments of tech jargon aliens give humans are neither self-validating nor useful

MUFON's "Star Team" is not, however, the first attempt within organized UFOlogy to create a "rapid response team" to quickly investigate reports. In an article in *Playboy* magazine way back in December, 1967, Hynek proposed (and later implemented) a national toll-free UFO Hotline, to be "manned 24 hours a day by competent interrogators capable of recognizing a true UFO report from a prankster's report.... If the report passes preliminary and immediate screening, headquarters notifies the local police and they rush to the scene." He explained how he expected solid and irrefutable UFO data "within a year of the initiation of such a no-nonsense program." But in a moment of perhaps unguarded optimism, Hynek added, "if the UFO-1000 program is sincerely and intensively carried out for a full year and yields nothing, this, in itself, would be of great negative significance. Then we could go back to the 'real, common-sense world' of pre-UFO days – shrugging it all off with 'There must have been a virus going around.' "

In an interview in *Saga UFO Report*, August, 1976, Hynek explained how his national hotline was working out: "In an unprecedented move, the FBI printed an article of mine in their monthly bulletin [February, 1975]. We furnished them with a special toll-free number which they can call 24 hours a day, seven days a week. Every night we get at least one call... we contact one of our 300 regional representatives, and they go and interview the witnesses." Geiger counters, soil samples, physiological effects, etc. are all involved in the investigation. Hynek gave no explanation of why he had not given up on UFOs as he earlier said he would if a year-long study yielded no solid evidence.

Other "rapid response" efforts to catch UFOs have likewise been attempted. Peter Davenport's National UFO Reporting Center has been collecting UFO reports on its telephone hotline since 1974, many from law enforcement and emergency service agencies, yet UFO proof continues to elude them. In 1977 France's CNES, their equivalent of NASA, created an agency GEPAN to officially sponsor investigations of UFO reports. It, too, failed to come up with anything really convincing, and CNES terminated all UFO investigations in 2004. In the 1990s, when according to news reports Mexico City was being inundated by a Saucer Blitz, Mexican UFO promoter and TV personality Jaime Maussan organized *Los Vigilantes*, who were supposed to be ready to respond to saucer reports with cameras and equipment at very short notice. They never obtained anything of significance.

At the 2015 MUFON Symposium, **Marc D'Antonio**, MUFON's chief photo/video analyst, described an ongoing effort to develop a multiple sensor device called **UFOTOG II** that will look for anomalous objects not only visually, but with magnetometers, gravity meters, spectroscopy, gamma ray and other detectors. Working with motion picture special effects guru Doug Trumbull, the plan is for large numbers of these devices to be manufactured to get the per-unit costs down, and place them on top of poles in areas where UFOs are being reported. Communicating via cell phone towers, their detection of radiation or magnetic events associated with anomalous objects might confirm that UFOs are "punching in" or out of other dimensions, as reportedly permitted by String Theory. (But more likely, they will not detect any such thing.) And **Mark Rodeghier**, Ph.D., who is the scientific director and president of the J. Allen Hynek Center for UFO Studies, is heading up a team to develop **UFODATA**, a UFO detection and tracking device. It is to be "a large network of automated surveillance stations with sophisticated sensors that will monitor the skies 24/7 looking for aerial anomalies," relying heavily on spectroscopy. *If* they can meet their crowd-sourcing financial targets.

On December 1, 2008 the British Ministry of Defence confirmed the closing of its so-called "UFO hotline" where individuals could call to report a sighting to the Ministry, and an associated email box. 135 reports were received during 2008. By doing so it expects to save the equivalent of about $73,000 each year, a trifling sum in the context of military budgets. There was only one man working in the department that looks into UFOs,

and he only part-time. UFO promoter Nick Pope had earlier held that position. A spokesman for the Ministry stated, "Our resources are focused on the priority – the front line in Afghanistan. Any legitimate threat to the UK air space will be spotted by our 24/7 radar checks and will be dealt with by RAF fighter aircraft." Pope told the Associated Press, "It's a great shame. This is the end of over 50 years of research and investigation into one of the biggest mysteries of our time." The MoD, however, stated, "In over 50 years, no UFO report has revealed any evidence of a potential threat to the United Kingdom...The MoD has no specific capability for identifying the nature of such sightings. There is no defence benefit in such investigation and it would be an inappropriate use of defence resources." The people of the U.K. need not be concerned that valuable UFO information might be lost by this decision, as no group in any country has succeeded, despite a number of attempts, in catching any UFO *in flagrante delicto,* even when there is a "hotline" for people to call at the very moment they believe they are seeing an alien spaceship in the sky.

Obviously Bigelow and MUFON must have expected that their "rapid response" efforts will bear more fruits than did these others. But for whatever reason, MUFON's arrangement with Bigelow did not turn out well. Some expressed the feeling that Bigelow sought to "purchase" MUFON's UFO database and other expertise. Worse yet, some charge that by "selling" its files, MUFON has violated privacy guarantees given to supposed "UFO abductees" who underwent hypnosis by MUFON hypnotists. John Schuessler writes on the MUFON website that "Unfortunately, dissident UFO buffs quickly came up with nonsense conspiracy theories about the cooperative agreement [with Bigelow] and spread malcontent and disinformation about it across the Internet." He does not say exactly how it ended.

## Promotion of UFO Belief Today

UFOs, like ghosts, ESP, Bigfoot, etc., are 'jealous phenomena': the alleged phenomenon invariably slips away before the evidence becomes too convincing [Sheaffer, 1998, p.211-222]. It is not the duration or location of the phenomenon limiting our ability to pin it down, but rather an evasiveness and a strange subjective visibility seemingly inherent in the phenomenon itself. Given the large numbers claimed for UFO abductions and close encounters, it is utterly incompatible with the laws of chance

that any such large-scale aerial operation could be taking place in major populated areas without leaving behind unambiguous evidence of its reality.

Initially, UFO excitement and belief was spread by news reports over mass sightings—a phenomenon that no longer occurs. Major magazines and books made sensational claims about sightings, which would generate much follow-on publicity. Pro-UFO books by Major Donald E. Keyhoe, Frank Edwards, John G. Fuller, and others became bestsellers and generated much interest and discussion of their claims. UFO groups such as NICAP and APRO appeared often in news stories about UFOs and were depicted as authoritative (rather than as groups devoted to promoting the idea of UFOs as interplanetary visitors).

Only rarely does a UFO book become a bestseller and generate nationwide interest and controversy. Whitley Strieber's *Communion* (1987) and *Transformation* (1988) were hugely successful books, written by a leading paranormal fiction author who swears that the astonishing events he is describing, unwitnessed and unverified, are nonetheless absolutely true. Today, first and foremost, the entertainment media play a major role in keeping UFOs alive, as well as radio and TV talk shows. News programs play only a very minor role. In the 1990s, cable-TV stations began producing entertainment programs based on popular UFO claims and themes, such as Roswell, the Alien Autopsy, and UFO abductions. In 2002, the Science Fiction (SyFi) channel presented Steven Spielberg's *Taken*, a twenty-hour miniseries based on alleged UFO abductions. Soon there were many other entertainment programs and movies featuring UFO themes, which even though presented as fiction many take to be "based on fact." Entertainment shows were soon bolstered by pro-UFO documentaries in which the skeptical view is given little or no voice and then by UFO "reality shows," such as The History Channel's *UFO Hunters* (2008), or the National Geographic Channel's *Chasing UFOs* (2012). In such shows, several "UFO experts" investigate UFO claims and invariably find tantalizing evidence yet never any real proof.

Talk shows on radio and TV also reach millions of people with their sensational claims and uncritical analyses. Since the 1980s, the syndicated late-night, call-in radio show *Coast To Coast AM* has reached millions— now on over 500 stations as well as on XM satellite radio. Originally

hosted by Art Bell, and now by George Noory, the show offers a dazzling array of wild tales about not only UFOs but cryptozoology, parapsychology, and conspiracies of every sort. Callers often relate their own allegedly paranormal experiences, and it seems that no claim is too bizarre to be given a respectful hearing.

Even some of the biggest names in the broadcast industry have uncritically promoted UFO claims in an attempt to boost ratings. During the summer of 2008, *Larry King Live on CNN* ran a series of poorly balanced programs about UFOs that displayed shockingly low standards of critical thinking for a major journalist. In February 2005, ABC-TV ran in prime time a two-hour show, "Peter Jennings Reporting: UFOs—Seeing Is Believing." The late journalist, a former news anchorman for ABC News, said, "I began this project with a healthy dose of skepticism and as open a mind as possible. After almost 150 interviews with scientists, investigators, and with many of those who claim to have witnessed unidentified flying objects, there are important questions that have not been completely answered—and a great deal not fully explained." In spite of all the reporting and investigative resources that must have been available to Jennings, the program contained nothing significant that had not already been reported before, and was just a re-hash on primetime network TV of existing UFO claims and interviews with mostly pro-UFOlogists.

# 2. SIGHTINGS OF UFOS

In the beginning there were sightings, and those sightings began with private pilot **Kenneth Arnold** (1915-1984) near **Mt. Rainer**, Washington on June 24, 1947. As soon as news stories appeared reporting Arnold's claim that he saw nine airborne objects that flew "like a saucer if you skip it across the water," others began reporting seeing the "saucers" too (a curious development, since Arnold did not say that the objects looked like saucers—they looked like boomerangs, he said—but skipped like saucers, a subtlety lost in the public's imagination). What did Arnold see? Over the decades many explanations have been given. In 1948, U.S. Air Force

scientific consultant Dr. J. Allen Hynek noted inconsistencies in Arnold's estimates, and suggested that he saw a formation of aircraft much closer than he thought, and hence moving much more slowly. The noted astronomer **Donald H. Menzel** (1901-1976) suggested **Arnold** saw meteor-ologically caused **mirages** of mountain tops in his 1953 book

*Kenneth Arnold with a drawing of one of the objects he reportedly saw in 1947 - looking frankly rather bird-like.*

*Flying Saucers*, but in the 1977 book he co-authored with Ernest Taves, *The UFO Enigma*, this was changed to suggest he saw water droplets on the aircraft window (!). In his **Skeptics UFO Newsletter** (SUN) #46 of July 1997, **Philip J. Klass** explains why he thinks **Arnold** saw a brilliant daylight meteor breaking up. In 1997 Martin **Kottmeyer** suggested that **Arnold** saw a flock of swans, which in 1999 UFO researcher

James **Easton** modified to a formation of American White **Pelicans**, the largest birds in North America.

This makes a lot of sense, as Arnold says he initially thought he was seeing a flock of geese. He also once compared the objects to a flock of blackbirds. He said the formation flew with a peculiar weaving motion 'like the tail of a Chinese kite.' The boomerang-shaped objects he drew actually looked quite bird-like. But many UFOlogists ridiculed this suggestion, and even coined the new pejorative label "Pelicanist," meaning one who crafts impossible explanations for UFO sightings. My **debunker.com** website has more information on Kenneth **Arnold**. In 2010 the Scottish UFO investigator Martin **Shough** wrote a long and exceedingly detailed investigation of Arnold's "**Singular Adventure**," concluding that none of the explanations offered are valid.

In the years following his initial sighting, Arnold said many things that are hard to believe. In an interview with reporter **Bob Pratt, Arnold** said says he has seen saucers seven or eight times since that first sighting (he said of his sighting, "I think this is probably the greatest discovery in the world"). "Once I saw a UFO which changed its density, so I concluded these things could be something alive rather than machines." He also claimed that his phone was being tapped, and that the saucers seem to be able to read his mind. We see that Arnold's claims can be rather fanciful and need to be taken with a big grain of salt. As for me, my money is on the **Pelicans**.

Soon after Arnold, sightings of "saucers" were pouring in from all around the country and from around the world. Sightings occurred in waves, which appeared to be fueled by media reports. A wave would typically start in one location, but as soon as news reports began to carry the story of the localized excitement, sightings activity would pick up nationally. Great waves of UFO sightings occurred in 1947, 1949, 1952, 1957, 1965—67, and 1973.

With the benefit of hindsight, we now know that the last large-scale national wave of UFO sightings occurred in the fall of 1973. The reasons for this are not clear. One common-sense explanation is that after more than twenty-five years of sensational sighting reports ultimately leading to nothing tangible and no new evidence, the public's fascination with saucer sightings was wearing out. One prominent UFOlogist, Karl Pflock

(1943-2006), later suggested quite seriously that extraterrestrial visitors actually did arrive around 1947 but departed sometime after 1973, and hence all subsequent UFO sightings were bogus. My preferred explanation is that this was right around the time that the majority of U.S. homes acquired color television, resulting in fewer eyes directed skyward, as well as the ennui factor. Whatever the reason, the "buzz" was gone for mass waves of saucer sightings. Individual and even localized clumps of sightings continued to occur and to be reported in the news, but somehow they were no longer contagious. "Lights in the sky" no longer were a "shiny new thing," and the public required something else to generate excitement about UFOs.

## UFO Sightings – Blips, but not Waves

Today, UFO sightings come in "Blips" instead of "Waves." Some of the Blips may well result in a major UFO case, like the famous Phoenix Lights of 1997. They may even repeat a few times But they do not become national in scope, and are not "contagious" in the way that UFO sightings used to be. No longer is an entire area set aflame with UFO excitement, as was Michigan in 1966, causing Venus and airplanes to be reported as UFOs over and over. No longer does any dramatic UFO event cause UFO excitement, and hence sightings, to go national, and persist for weeks or months. Nowadays UFOlogists have fewer sightings with which to build their case, and so must choose them more carefully.

## A "Top Ten" UFO Case – Yukon, Canada, 1996

"We are all greater artists than we realize," Nietzsche famously wrote, describing the way our minds construct reality. The following incident illustrates the "artistry" to which Nietzsche referred.

On the evening of December 11, 1996, more than 30 people in several different locations in Canada's sparsely populated Yukon Territories reported seeing a huge "UFO mothership" with rows of lights, flying by as a Close Encounter of the First Kind.

The documentary film *Best Evidence: Top 10 UFO Sightings* by Paul Kimball, available on *YouTube*, lists this "multiple witness sighting in the Yukon" as number eight of the top ten UFO cases of all time. In that film the celebrated "Flying Saucer Physicist" Stanton Friedman says:

"The Yukon case *is* emblematic of what a good case should be. I mean, sure, we'd like to have a piece of the craft, we'd like to have the crewmember introduced for dinner. *But* multiple independent witnesses lasting a long time, describing something that's *way* outside the norm, -- there's no way you can make it into a 747, for example [chuckle]. And big, but this was much much bigger than a 747. "

Longtime UFOlogist Michael Swords of the Center for UFO Studies says:

Not knowing [investigator] Martin Jasek I can't "stand up in court" on this one, but everything that I've heard says that this is not only a "good" but possibly one of the best cases ever... I look forward to any of the gang clearing my misconceptions up on this case, because right now it might be one I'd "take into war" with me.

## What Was It?

In 2012, skeptic James Oberg contacted the Canadian satellite expert Ted Molczan with the details of this case. Molczan is probably the world's top civilian expert on observing earth satellites and calculating satellite orbits. **Molczan** looked into the matter carefully, and came up with an exact match: "the observed phenomena were due to the re-entry of the 2nd stage of the rocket that placed **Cosmos 2335** into orbit earlier the same day" (NORAD designation 1996-069B / 24671). Should anyone doubt this, Molczan provided details of the mathematical calculations that support this conclusion. Later, satellite expert Harro Zimmer refined the details of the re-entry decay of the Russian Cosmos 2335 rocket booster, giving even greater precision to the object's position and velocity during re-entry. It fully confirms the identification Molczan made earlier. The world's experts on satellite orbits and re-entries are in full agreement: the 1996 Yukon UFO coincides exactly with the breakup of the rocket booster of Cosmos 2335.

Oberg placed a comment on the *Above Top Secret* forum discussing this case, to which many objections were, of course, made. But this is not the first time a satellite re-entry has given rise to widespread UFO reports. There was the famous **Zond 4** re-entry **UFO** in **1968,** re-entries seen widely in Europe in 1990 and 1994, and others that Molczan has been working diligently to identify. In fact, Molczan has created a dedicated web page listing all known visually-observed natural re-entries of earth satellites, which will be updated as necessary: http://goo.gl/TD8FWn Reports of UFOs can be compared to this list to see if the incident was caused by a satellite falling out of orbit. Of course, if the incident was caused by a brilliant meteor fireball, it will not be on this list.

# Stimulus / Response

A case of this type affords us an excellent opportunity to judge the credibility of eyewitness testimony. Given a known stimulus "in," what is the observer's response "out"? In other words, how accurately did the observers' descriptions match the known stimulus? Not well at all!

*Report:* "many rows of lights"

*Reality:* The booster disintegrated into an irregular train of debris that was perceived as an orderly pattern of "lights" on a huge solid object.

*Report:* "As he was walking his flashlight happened to point in the direction of the UFO. As if reacting to his flashlight, the UFO started speeding rapidly toward him."

*Reality:* the "UFO reacting" to him was entirely in his imagination. The rocket booster did not react to his flashlight.

*Report:* the UFO was hovering approximately 300 yards in front of the observer. "Hynek Classification: CE1" (Close Encounter of the First Kind).

*Reality:* the distance to the re-entering booster was approximately 233 km (145 miles), so this was not a "close encounter." At no time did it stop, or hover.

*Report:* The UFO was approximately 500-750 meters (up to 1/2 mile) in length.

*Reality:* It is impossible to estimate the size of an unknown object unless its distance is known. Since the disintegrating booster was about 145 miles distant, its debris train must have been spread over many miles.

Report: "The interior lights in her car started to go dim and the music from her tape deck slowed down."

*Reality:* This effect was entirely in the observer's imagination. The rocket booster did not affect her car's electronics.

*Report:* "stars blocked out" by huge UFO.

*Reality:* the observers were viewing a long train of debris from the disintegrating rocket booster. It was not a solid object, and thus could not have "blocked out" stars. However, the light from the reentry may have made nearby stars difficult to see.

Many UFO proponents criticize the above analysis as absurd. One anonymous poster said, it "ignores 90% of witness testimony then discounts the rest." But this places them in the position of claiming that their alien spaceship was seen flying at exactly the same location and time and in the same path where the flaming rocket booster was re-entering the atmosphere. Molczan closed his analysis by saying,

Experienced sky watchers on SeeSat-L may find it difficult to believe that anyone could misidentify a re-entry as a spaceship, but human perception is notoriously fallible, and no one is immune. Much depends on the circumstances and personal experience. Driving through the wilderness under a pitch black sky, and suddenly faced with a slowly moving formation of brilliant lights can be awe-inspiring and even terrifying. The human mind races to make sense of the unfamiliar, drawing on experience that may be inadequate. Depth perception can play tricks, such that something 200 km away, 100 km long, and moving at 7 km/s, seems to be just 200 m away, 100 m long, and moving 7 km/h - the angular velocity is roughly the same. Taking these considerations into account, the eyewitnesses did a pretty good job, and need not be embarrassed for having perceived more than was there.

He left out the part where the observers reported object hovering, the electrical interference, etc. So I wouldn't call that "a pretty good job" of observing.

Here we have yet another clear-cut example of extraordinary reports ("giant UFO Mothership!") arising from a perfectly ordinary (if rare) phenomenon. ***Therefore, the existence of extraordinary reports does not suggest the existence of extraordinary objects.*** It is perfectly possible to get extraordinary reports from ordinary objects

## 1986: Captain Terauchi's Marvellous "Spaceship"

The San Francisco Chronicle reported on December 30, 1986, "The crew of a Japan Air Lines cargo jet claimed that a UFO with flashing white and yellow strobe lights followed them across the Arctic Circle in route from Reykjavik, Iceland, to Tokyo." On January 1, 1987, that paper reported, "A veteran pilot whose UFO sighting was confirmed on radar screens said the thing was so enormous that his Japan Air Lines cargo jet - a Boeing 747 - was tiny compared with the mysterious object." In fact, Terauchi said that the object was larger than an "aircraft carrier." Feeling the heat, the FAA soon re-opened its investigation of the incident. "The reason we're exploring it is that it was a violation of airspace," said FAA spokesman Paul Steucke. "That may sound strange, but that's what it was."

The FAA reviewed its data, and found reasons to doubt its earlier statements. By Jan 8, the press was reporting,

> The FAA has concluded that the unidentified object on radar now appears to be an unexplained split image of the JAL Boeing 747 and not a separate object .... The review of radar data indicates that no second object was present and represents a reversal of earlier FAA statements that a second object was confirmed on radar. "The bottom line is that this tells us that we don't have any radar confirmation of the object that the pilot said he saw."

Skeptic Philip J. Klass investigated, and soon CSICOP issued a Press Release, written by Klass:

# Sightings of UFOs

At the time the UFO incident began near Ft. Yukon, the JAL airliner was flying south in twilight conditions so that an extremely bright Jupiter (-2.6 magnitude) would have been visible on the pilot's left-hand side, where he first reported seeing the UFO, according to Klass. Jupiter was only 10 degrees above the horizon, making it appear to the pilot to be at roughly his own 35,000 ft. altitude. Mars, slightly lower on the horizon, was about 20 degrees to the right of Jupiter but not as bright....Although the very bright Jupiter, and less bright Mars, had to be visible to JAL Capt. Kenjyu Terauchi, the pilot never once reported seeing either -- only a UFO

Many of the colorful details of the incident carried by the news media, largely based on the six-week-old recollections of the pilot of JAL Flight 1628, are contradicted by a transcript of radio messages from the pilot to FAA controllers while the incident was in progress. For example, news media accounts quoting the 747 pilot said that when he executed a 360 degree turn, the UFO had followed him around the turn. But this claim is contrary to what the pilot told FAA controllers at the time.

Klass wrote an article in *The Skeptical Inquirer*, Summer 1987: "FAA Data Sheds New Light on JAL Pilot's UFO Report." Klass wrote,

The FAA data package reveals Terauchi to be a "UFO repeater," with two other UFO sightings prior to November 17, and two more this past January, which normally raises a "caution flag" for experienced UFO investigators. The JAL pilot is convinced that UFOs are extraterrestrial and when describing the light(s) Terauchi often used the term *spaceship* or *mothership*.

[Terauchi] always failed to mention that two other aircraft in the area that were vectored into the vicinity of the JAL 747 to try to spot the UFO he had been reporting were unable to see any such object... [Flight Engineer Yoshio Tsukuba] "was not sure whether the object was a UFO or not"... When the copilot [Takanori Tamefuji] was asked if he could distinguish these lights "as being different" from a star, he replied: "No."

Re-reading Terauchi's own statements about the incident, I don't think that anyone could call him an unbiased or objective observer.

# The Famous "Phoenix Lights" – March 13, 1997

The most famous UFO event of recent years was the celebrated Phoenix Lights, on the evening of March 13, 1997. A lot of people were outdoors that evening, trying to get a view of the celebrated bright comet Hale-Bopp that was then well placed for observation. The Phoenix Lights actually consisted of two separate events, the first beginning approximately 8:00 PM and ending before 9:00, and the second one starting approximately 10:00 PM. For purposes of analysis, the two events must be considered separately, as they have different causes. UFO proponents, however, love to jumble the two together, because it makes for a more dramatic case. Typically, UFO TV programs will show you over and over again the visual of the second incident, which is dramatic but easily explained, while they are discussing the first incident, for which there is much less photographic evidence.

## The First Incident – a Flyover by Lights in a V-shaped Pattern

The one thing that everyone can agree on is that starting around 8:00 PM Mountain Standard Time a sequence of lights in a V-shaped formation was seen flying southeastward across much of the state of Arizona, from Interstate 40 near Prescott in the north to Phoenix and Casa Grande in the south. The earliest sighting was reported at 7:55 PM MST by a young man in **Henderson**, Nevada, just southeast of Las Vegas. He informed the **National UFO Reporting Center** of seeing "a **V-shaped object**, with six large lights on its leading edge, approach his position from the northwest and pass overhead." The marathon UFO evening continued when observers in the northern part of Arizona, looking toward the comet in the northwest, saw five bright lights in a V-shaped formation moving toward them. Some described it as a huge, solid V-shaped object that blocked out the stars as it passed over. Others described it as five unconnected lights. However, the only **video** in existence of the first incident, is that of **Terry Proctor**, and in its 43-second length clearly shows the motion of the objects with respect to each other. In other words, despite some observers' impressions, what flew over was not a single solid object, but five unconnected lights flying in formation.

# Sightings of UFOs

As the lights passed over the Phoenix area they were seen by hundreds of people. The investigative reporter **Tony Ortega** wrote in **The Great UFO Cover-Up** in the *Phoenix New Times* (June 26, 1997) that one important observer's account had been ignored by almost everyone: Scottsdale amateur astronomer Mitch Stanley, then 21, who was outside observing with his ten-inch reflecting telescope. Because the Dobsonian mount of his telescope moves flexibly and freely, he was able to examine the lights under magnification. He clearly saw that they were airplanes: "What looked like individual lights to the naked eye actually split into two under the resolving power of the telescope. The lights were located on the undersides of squarish wings, Mitch says." Stanley tried to give his account to reporters and local politicians who were making big deal out of the sightings, but nobody paid any attention to him, until he finally spoke to Ortega. Later UFO skeptic James McGaha interviewed Stanley at length. Stanley was not familiar with the appearance of the A-10 aircraft, but the description he gave left McGaha with no doubt that he had seen a flight of A-10 Warthogs, a plane that McGaha became very familiar with while he was in the Air Force. These aircraft are used by Air National Guard units across the U.S.

UFO Skeptic **Tim Printy** gives an excellent explanation of the **V-Formation** of lights leaving little doubt that these were aircraft in formation. He found and interviewed yet another observer whose account has been ignored by UFO proponents. Rich Contry was driving west that evening on I-40 north of Prescott, and observed the objects in 10x50 binoculars:

> I was on my way from Flagstaff to Laughlin Thursday when I saw the light formation reported on the radio the other night. I'm a pilot and was in the U.S. Air Force 4 years. Being in the mountains on highway 40, the night was clear and still. As the formation came towards me I stopped my car and got out with my binocs to check out what this was. As it came towards me, I saw 5 aircraft with their running lights (red and green) and the landing lights (white) on. They were also flying fairly slow and in the delta formation. As they went over me I could see stars going between the aircraft so it could not have been one large ship. The flying was like that of the Blue Angels or the thunderbirds demo team. Also as they went by their jets were not very loud because of the low throttle setting for flying slow but I did hear the jets as they went away towards the south.

Bad UFOs

McGaha notes that the A-10 is the quietest jet that the Air Force has, so it is not surprising that many witnesses heard nothing.

Printy also interviewed Mitch Stanley, who explained that he had been using a Televue 32mm Plossl eyepiece that gives a magnification of 43x. The telescope's field of view would have been a little over one degree, or twice the size of the full moon. Printy writes,

> He states that in addition to the lights on the wings, he also saw one underneath the fuselage, '3 lights! I saw 3. One on each wing and one on the fuselage' (Stanley). This confirms Rich's description to me that one of the lights was illuminating the front landing gear. [UFO chronicler] Peter Davenport also confirms that the lights were possibly composed of two or more lights as Mitch states, 'Mitch is correct that each of the larger lights was, by at least two good observers, reported to consist of two or three individual lights.'

So at this point, there is no real doubt that the V-shaped lights were a formation of terrestrial aircraft. The only remaining questions are, what kind of aircraft were they, who was flying them, and where were they flying? According to a May 1999 **Reader's Digest** article by Randy Fitzgerald, **UFOs – A Second Look**,

> At 8:30 p.m. the cockpit crew of an American West 757 airliner at 17,000 feet near Lake Pleasant, Ariz., noticed the lights off to their right and just above them.

> "There's a UFO!" co-pilot John Middleton said kiddingly to pilot Larry Campbell. They queried the regional air-traffic-control center in Albuquerque, N.M. A controller radioed back that it was a formation of CT-144s flying at 19,000 feet.

> Overhearing the exchange, someone claiming to be a pilot in the formation radioed Middleton. "We're Canadian Snowbirds flying Tutors," a man said... [Fitzgerald later added] "We're headed to Davis-Monthan Air Force Base."

"Snowbirds" is the name of the Canadian Flying Demonstration team, similar to the Blue Angels. But that story does not check out – the Canadian Snowbirds were not in Arizona at that time. So what does that mean? Why would a pilot, apparently flying quite legally in a formation of five aircraft, misidentify his aircraft and his mission? **Fitzgerald** has gone so far as to suggest in a 2010 two-part *Examiner* article that the apparently deceptive reference to the Canadian **Snowbirds** may have been an exercise in "**psychological warfare**." He notes that "When the first UFO sighting reports began coming in that night, the formation of lights were seen coming from the direction of Las Vegas along a commercially trafficked air corridor." He suggests that this might have been some secret exercise involving technology developed at Area 51 near Las Vegas, possibly including "Holographic Deception Technology."

Since there is now no recording of that cockpit conversation, we cannot be sure of the exact words used. But suppose that Middleton, repeating the conversation to Fitzgerald, misremembered the exact words? Suppose it went something like this: "We're Snowbirds, we're headed to Davis-Monthan Air Force Base." Middleton then reflected, "Snowbirds, they're the Canadian Flying Demonstration team." Because in that case, he would have made a very understandable mistake: he confused two different flying groups that use the same name!

*Since the 1970s, Davis-Monthan Air Force Base in Tucson has operated a program called Operation Snowbird, which brings in Air National Guard pilots from snowy northern states for winter flying practice.* It is operated by the Air National Guard's 162nd Fighter Wing. In recent years, foreign pilots also have received training there, and some Tucson residents are very unhappy with *Operation Snowbird* because of the large number of jets taking off and landing.

Fitzgerald notes in his *Examiner* article that "Once beyond the southern suburbs of Phoenix, the formation of lights followed Interstate 10 toward Tucson." This is strange behavior for an Extraterrestrial Craft. One man told Fitzgerald, "This was so profoundly my most significant visual experience ever, like the hand of God coming down." There were several more observers along Interstate 10 as the objects came down lower and approached Tucson, who described them as five separate objects, and not one huge one. The time was noted as 8:42 PM. These were the last sightings of the V-shaped lights.

March 13, 1997 was the last night of *Operation Snowbird* (Scott Craven, *Arizona Republic*, Feb. 25, 2007), so obviously any aircraft that had flown off to other destinations, such as Las Vegas, presumably had to return to Tucson. James McGaha reconstructs the story of the V-shaped formation as follows: Five A-10 jets from *Operation Snowbird* had flown from Tucson to Nellis Air force Base near Las Vegas several days earlier, and were now returning. The A-10 jets were flying VFR (visual flight rules), so there was no need for them to check in with airports along the route. They were following the main air corridor for air traffic traveling that route, the "highway in the sky." (Why a UFO would follow U.S. air traffic corridors is a mystery.) Because they were flying in formation mode they did not have on their familiar blinking collision lights, but instead their formation lights. In any case, FAA rules concerning aircraft lights and flight altitudes, etc. do not apply to military aircraft. The A-10s flew over the Phoenix area, flew on to Tucson, and landed at Davis-Monthan.

When the U.S. Air Force was queried about both incidents, their (correct) initial response was that they knew nothing about them. Both UFO incidents that night involved aircraft of *Operation Snowbird*, and the Air National Guard has a completely separate command structure from the Air Force. Each seems to know surprisingly little about the other! Unfortunately, by the time that all of this was pieced together, all records of routine flight operations had been destroyed. But there is very little room for doubt that the famous "V-shaped formation" of lights was five Air National Guard A-10 aircraft flying into Davis-Monthan Air Force Base in Tucson.

## The Second Incident – a Flare Drop

Starting around 10:00 PM that same evening, hundreds if not thousands of people in the Phoenix area witnessed a row of brilliant lights hovering in the sky, or slowly falling. Many photographs and videos were taken, making this perhaps the most widely-witnessed UFO event in history. As explained in the **Wikipedia** article on the **Phoenix Lights**,

> The U.S. Air Force explained the second event as slow-falling, long-burning LUU-2B/B illumination flares dropped by a flight of four A-10 Warthog aircraft on a training exercise at the Barry Goldwater Range at Luke Air Force Base. According to this explanation, the flares would have been visible in Phoenix and appeared to hover due to rising heat from the burning flares creating a "balloon" effect on their parachutes, which slowed the descent. The lights then

appeared to wink out as they fell behind the Sierra Estrella, a mountain range to the southwest of Phoenix.

A Maryland Air National Guard pilot, Lt. Col. Ed Jones, responding to a March 2007 media query, confirmed that he had flown one of the aircraft in the formation that dropped flares on the night in question. The squadron to which he belonged was in fact at Davis-Monthan AFB, Arizona on a training exercise at the time and flew training sorties to the Barry Goldwater Range on the night in question, according to the Maryland Air National Guard. A history of the Maryland Air National Guard published in 2000 asserted that the squadron, the 104th Fighter Squadron, was responsible for the incident. The first reports that members of the Maryland Air National Guard were responsible for the incident were published in *The Arizona Republic* newspaper in July 1997.

Many UFO proponents, including MUFON's State Director for Arizona, Thomas R. Taylor, accept that that the video of the second incident - bright lights in a row - was in fact a flare drop, while maintaining that the first incident is still unexplained.

Former Arizona governor Fife Symington now claims to have 'held back' UFO information, and claims to have seen the V-shaped UFO flying over, although he admits he didn't tell anyone about this for ten years.

*Symington is lying.* In an interview on **Larry King Live** on CNN (July 13, 2007), **Symington** said:

> I was up in the sunny slope area around 8:00 at night. And I went out to look to the west where the -- all the news channels were filming the Phoenix Lights. And to my astonishment this large sort of delta-shaped, wedge-shaped, craft moved silently over the valley, over Squall Peak, dramatically large, very distinctive leading edge with some enormous lights. And it just went on down to the Southeast Valley. And I was absolutely stunned because I was turning to the west looking for the distant Phoenix Lights and all of a sudden this apparition appears.

There is no way this could have happened around 8:00 PM, because Symington says that news channels were already filming the Phoenix

Lights in the west – the flare drop. We know that the flare drop did not begin until about 9:57 PM, and surely if news crews were already out photographing them, it must have been well after 10:00. In fact, it was probably closer to 11:00 PM, because Symington said he was "looking for the distant Phoenix Lights," implying that the 'show' may have been over. This is when he claimed to have seen the V-shaped object make a dramatic flyover. All the other observers, however, agree that the V-shaped lights were not seen after about 8:45 PM. If Symington claims that he saw a giant V fly over after 9:00 PM – and probably after 10:00 PM – then he is the only one who did.

How credible is Fife Symington? The *New York Times* news story of Sepember 4, 1997 tells you all you need to know: **Arizona Governor Convicted Of Fraud and Will Step Down**. "A Federal jury today convicted Gov. Fife Symington of Arizona of seven felony counts of defrauding his lenders as a commercial real estate developer, forcing his resignation and leaving him facing the prospect of prison and hundreds of thousands of dollars in fines." Sure, I believe everything this guy says. (Symington's conviction was later overturned on a technicality. Before he could be retried, he was pardoned by the outgoing President Clinton, so Symington got off scott-free.) What might be his motivation for making up this UFO Yarn? An *Arizona Republic* article of September 3, 2007 is headlined, "**Ex-Governor Pursues New Ventures; Beset By Scandal**, Now He Enjoys Life As Entrepreneur." It notes, "Unlike many others who have been brought down by scandal, Symington didn't disappear from the public radar screen. In recent months, he has been on national television twice, commenting about a longtime interest: lights that appeared over Phoenix in 1997." Think of how much Symington would have had to pay to get the publicity and name recognition that he got for free by making up his UFO tale.

But Dr. Lynne D. Kitei, M.D. isn't having any of this "flare drop" business. Her website *ThePhoenixLights.net* says it promotes "Evolution to a New Consciousness." She claims she was watching the Phoenix Lights two years before everyone else, and that her research proves "we are not alone." By some complicated analysis she claims to have proven that the objects could not have been flares, although I haven't run across anyone who understands it, and agrees she is correct. I heard her speak at the 2012 *International UFO Congress*, and some of her photos of UFOs appeared to me to be lights on the ground. Giving up her practice of medicine to become a full-time UFO promoter, "Dr. Lynne" (as she is sometimes fondly called) has made a documentary film *The Phoenix Lights*, and has appeared on *Coast to Coast AM,* the well-known late-night paranormal and conspiracy-fest, to tell her tales.

The famous Lights have returned to Phoenix on several occasions. An Air Force flare drop on February 6, 2007 gave rise to widespread UFO reports. And on April 21, 2008, a private individual apparently attached flares to his own helium balloons, and sent them up. He remained anonymous, which is not surprising since he could be held responsible for any fires or other damages his airborne flares might cause.

## Parachuting UFOs

On December 1, 2010, the Phoenix Lights appeared again, but this time in Los Angeles, California. And the primary witness was San Antonio Spurs star **Manu Ginobili**, who was in town to play the Clippers. Ginobili and some friends spotted glowing objects slowly descending in the sky. A **video** of the sighting was taken and posted it to Manu's Facebook page, where it attracted a lot of attention, and soon ended up on the website **TMZ**.com that deals in celebrity news

In another classic example of clueless "expert" commentary, the next day TMZ.com offered "UFO Expert -- On Board with NBA Star's Theory." UFOlogist Robert Kiviat, the executive producer of the now-discredited "Alien Autopsy (Fact or Fiction?)," was suggesting that "one strong possibility is the UFOs were experimental military craft." Kiviat believes the objects were part of "some sort of military test and farther away from the area than they appear," a conjecture without any facts to support it.

Finally, four days after the "UFO" was first posted, TMZ.com reported, "TMZ has uncovered the truth behind the UFOs he spotted over L.A. earlier this month ...and turns out, Red Bull is to blame!!!" As the website explains, "A rep for the company tells **TMZ** ...a team of **Red Bull Air Force** skydivers took the plunge over **Santa Monica** at around 5:00 PM on December 1 -- the exact time and day Manu was caught on tape watching glowing objects descend from the sky. As for the glowing? We're told the skydivers were carrying powerful flares during their descent." On the website for the **Red Bull Air Force**, parachutist **John de Vore** writes,

# Bad UFOs

I was watching the news and I see them reporting on 2 UFO sighting in **Santa Monica**. The sightings were on Dec 1 & 8. As soon as I saw the videos on the news I busted up laughing. It was us jumping with our night flares.

Mick West's excellent website **Contrailscience.com** has a full analysis of the **Ginobili** photos, including the use of Google Earth to show precisely where the objects were, above the beach in **Santa Monica**.

On a Friday evening, October 28, 2011, the action moved back to the Phoenix metropolitan area. Four bright lights in the sky were reported and photographed at a high school football game in Scottsdale, Arizona. A video of the supposedly unexplained lights were shown by broadcaster Mark Mancuso on Accuweather. The video apparently was soon pulled, probably because it made Accuweather a laughingstock, and put Mancuso in the woodshed. It was not hard to see the similarity to the parachuting flares of the Phoenix lights. SkyFOX helicopter pilot Rick Crabbs (Fox Channel 10 in Phoenix) said, "I was at the location where those skydivers were coming in ... Friday night, so that's exactly what happened -- there were some skydivers," he said. "And they did have pyrotechnics on their ankles. There were four of them, and if you look at the video, you can see four different lights." The skydivers were at an event called the "Halloween Balloon Spooktacular" at the Salt River Fields. Looking at the schedule of activities for that event, we find:

> 9:00pm Arizona Skyhawks performance with sky divers in lighted suits and pyrotechnics

The **Arizona Skyhawks** are a professional skydiving team that sometimes uses pyrotechnics to create a dazzline nighttime display. But for some "journalists," wild speculation is more gratifying than a little bit of actual research.

The "Phoenix Lights" repeated themselves yet again over Michigan in July 2012. Military exercises in the northern part of lower Michigan resulted in widespread UFO reports. Fortunately, this time it didn't take months or longer to establish what was happening. **Operation Northern Strike 2012** was the largest ever air-to-ground training exercise hosted in Michigan. According to the Air National Guard, **flare decoys** were being used to mimic air-to-ground missiles, and were easily visible from the ground.

## 2006 - A UFO Hovers over O'Hare Field

Relatively little has been written about the "UFO Incursion at O'Hare Airport" on Nov. 7, 2006, which is a major case for Leslie **Kean**, in her book *UFOs: Generals, Pilots, and Government Officials Go on the Record*. Several employees of United Airlines reported seeing a "strange object hovering just under a cloud bank... the metallic-looking disc was about the size of a quarter or half dollar held at arm's length." Unfortunately, no photographs exist of this supposed "metallic-looking disc" hovering over one of the world's busiest airports in daytime, and nothing showed up on radar. Even more surprising, we learn that the UFO hovered over gate C-17 at O'Hare. Apparently it was not seen by anyone at Gate C-15, C-16, or anywhere else. Supposedly the object hovered below the clouds at an altitude of 1,900 feet or less. If so, then from other vantages around the airport, it would have been seen not overhead, but at elevations of 45 degrees or less, making it possible to triangulate the object.

After approximately five to fifteen minutes, "the suspended disc suddenly shot up at an incredible speed and was gone in less than a second, leaving a crisp, cookie-cutter-like hole in the dense clouds. The opening was approximately the same size as the object [I would suggest that the

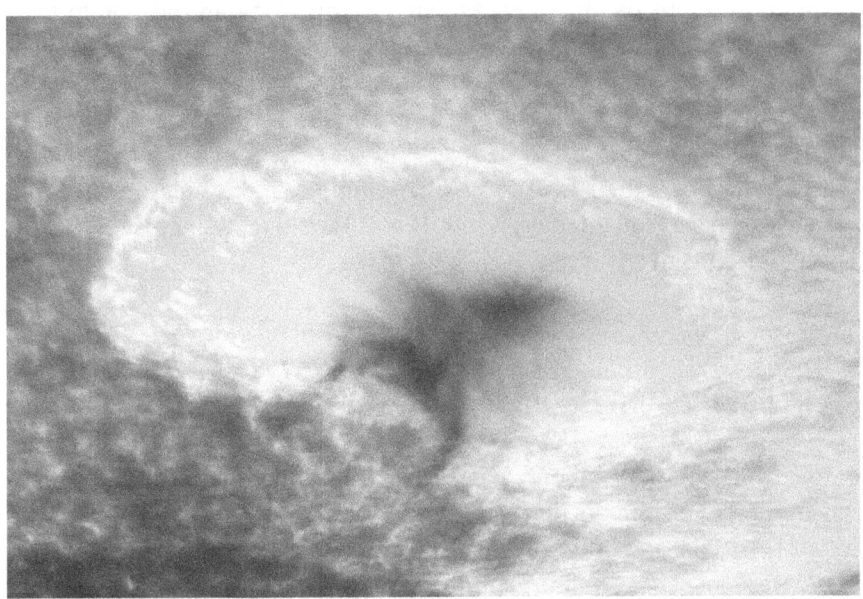

*A hole-punch cloud in Austria (Wikimedia Commons)*

opening *was* in fact the object], and those directly underneath it could see blue sky visible on the other side." Kean ridicules an explanation offered by an FAA spokesman that the observers saw a "hole-punch cloud," an unusual weather phenomenon where a large, dramatic circular hole is formed in a cloud layer. She cites a report by **NARCAP**, a pro-UFO investigative team, showing that temperatures were too high for a hole-punch cloud to form at the 1,900 foot elevation of the ceiling, which is probably correct. (Kean only references investigations by other researchers if their conclusions agree with hers.) But then she bizarrely suggests that "this just happens to fit the witnesses' explanation of what they saw: a high-energy, round object very likely to be emitting some form of intense radiation or heat while cutting through the cloud bank." Now, one cannot simultaneously argue that a hole-punch cloud could not have formed because the temperatures were above freezing, but a UFO formed one anyway. In any case, the low ceiling might easily have briefly parted to reveal a much higher cloud layer, where a hole-punch cloud already existed. NARCAP's anonymous witness "Ramp Agent X" said of the object, "it was almost the same color as the clouds and if you looked away it was hard to find it and focus again." This sounds very much like a hole-punch cloud. Another anonymous witness Mrs. J.H. supposedly was observing the object from a parking lot one mile distant. From a distance of one mile, an object 1,900 feet above the ground would be seen at an elevation of about twenty degrees, which is quite low in the sky. In fact, she claims to have initially seen the object at an even greater distance, probably two or three miles. Yet nobody claims to have seen it low in the sky, it was always overhead. She also remarks that it was difficult to see, and you had to point it out to people. It is interesting that the photo used on a **NOAA** website to illustrate the phenomenon of the **hole-punch cloud** was taken exactly eight days after the O'Hare Field "incursion," from nearby **Wisconsin**.

## Night Vision Equipment Reveals Fleets of UFOs

As an amateur astronomer participating in public outreach science education, I often encounter beliefs and questions about UFOs, extraterrestrials, and other unusual subjects. However, never before had I encountered such a well-organized, dedicated, but ultimately clueless group of UFO seekers as those who turned up at one of the monthly public star parties in 2010 conducted by the San Diego Astronomy Association just outside the Fleet Science Center in San Diego's Balboa Park. Five young adult men, joined later by several women, were showing off their new night vision binoculars and cameras, while asking all manner of questions about telescopes. It turns out that they were participants in one of UFOlogy's latest fads, using pricey high-technology night vision equipment to reveal all manner of things that the naked eye doesn't see – and calling what they don't understand "UFOs."

They were inspired by claims they had heard from a number of sources, including the *Coast to Coast AM* radio show. Night vision equipment reveals all manner of objects that are too faint to be seen by the unaided eye. If you see something you can't identify, then it must be a UFO. Sometimes they point a powerful green laser at such objects so others can try to see them. I warned them that if you are pointing a laser at an unknown aerial object, you just might be shining it into the cockpit of an aircraft. The point was appreciated. When it was my turn to look through the night-vision binoculars (which I believe sell for approximately $2,500), I was astonished at how much I could see. Standing under the light-polluted urban skies, with many nearby streetlights and with a marine layer developing (that's "fog", to non-Californians), one could hardly see stars fainter than about magnitude two (the stars in the Big Dipper). But using the night vision device, which interestingly did not magnify objects but simply amplified light, I could easily see all of the stars in the sickle of Leo, and the faint stars of Ursa Major, as if I were out under dark skies in the country. Suddenly, a very bright white object dashed through my field of view, covering the distance from zenith to about 20 degrees elevation in a few seconds. A flying saucer? Except that it banked and turned back in my direction, and I could detect the flapping of wings. An owl had been spotted several times circling a lighted building nearby. When that owl flew directly over us, away from the lights, it was too dark to be seen with the naked eye, but the goggles easily picked it up.

I was told that they had captured objects moving along like satellites that suddenly stopped and reversed themselves, which no satellite could do. However, none of the video clips they showed me on their camera confirmed that. They did show a clip of two satellites traveling side-by-side in tandem. I explained that they had very likely captured one of the clusters of the **Naval Ocean Surveillance System** satellites (NOSS). The U.S. Navy operates several clusters of two or three satellites that maintain a strict formation, that are apparently used to pinpoint the source of radio and radar transmissions. Normally they are too faint to be easily seen by the unaided eye, although occasionally the sun angle is such as to make them brighten to naked-eye visibility. When UFO seekers stumble across one of these clusters using optical magnification or amplification, they often conclude that they are seeing a formation of UFOs.

Night-Vision UFO watching is becoming a popular activity. The leader of this pack is Ed Grimsley, who claims to have videos of "Objects in Earth's space shooting it out," which he is happy to sell you. He even once had a Meetup Group in San Diego, where for a mere $20 you would be taken out to dark skies to see these amazing night vision UFOs yourself. What none of these intrepid explorers seem to realize is that the "mysterious" objects they thought were visible only in their green goggles were also visible to anyone who has a pair of binoculars - and the binoculars show the objects in better detail.

**Tim Printy** wrote some interesting articles "**Donning UFO Goggles**" and "**Night Vision Follies**" in his WebZine *Sunlite*. He notes that Ed Grimsley claims to have recorded "space battles" of UFOs, in which there are as many as 45 "kills" in one battle, which he is selling on DVDs. "Whoever it is we are battling, it is very serious and a threat to our National Security," Grimsley warns on his website. In fact, Grimsley claims to have been watching UFOs in the sky battling it out since he was a boy in 1961. Printy counters that it is "astonishing in that no astronomical telescopes or astronomers (professional or amateur) have seen these events even though Grimsley was able to see them with his unaided eye before he obtained his night vision goggles."

Printy continues, "Amateur and professional astronomers have been monitoring the skies with equipment far superior to what Grimsley is using for many years. For instance you can read the following links [*descriptions of astronomers' photographic sky survey programs*].... These articles have been written in the past decade but the footnotes demonstrate that people have been doing this kind of work for some time. Does Mr. Grimsley seriously believe that all of these astronomers are not seeing this or, is it possible that these astronomers see what he sees and can readily identify them?" Printy demonstrates that some of the alleged night-vision UFOs shown on DVD correspond to satellite passes.

The night-vision UFOlogists I encountered wanted to buy a telescope, and apparently were willing to spend $1,000 or more on it, to get close-up videos of these objects. They were thinking about getting a Cassegranian reflecting telescope as large as 8", which can easily deliver a useful magnification of about 400x under optimal conditions. (Cheap telescopes sold in department stores often lure the unsophisticated with promises of magnifications of 600 or more. However, any telescope can theoretically achieve any magnification you want by throwing additional Barlow lenses onto it, each one doubling the power – but the image quality quickly degrades and it becomes useless. Such telescopes will only perform satisfactorily at about 100x or less.) The UFOlogists could then demonstrate the existence of secret, off-the-record surveillance satellites, or possibly even actual extraterrestrial vehicles. After all, didn't some amateur photographers capture the ISS clearly enough to reveal its shape? Yes, that's true, I explained. But there exists software to track earth satellites using a telescope with a go-to mount, and the objects that you want to image are on unknown paths, so they must be tracked manually. The greater the magnification, the smaller the field of view, and hence the greater problem tracking. I suggested they look into a small, wide-angle telescope of 3 or 4 inches diameter, or better yet, large binoculars.

However, even if there were something spooky up there to track, this approach is most unlikely to yield high-resolution images. Night vision optics trade sensitivity for resolution, and bright objects spill out into a circle of light that in no way reflects their angular size. A good pair of binoculars will reveal fainter objects than its night-vision counterparts, albeit with a smaller field of view, and much better resolution. But now that the fad for imaging night vision UFOs has started, and sophisticated equipment is eagerly being promoted to unsophisticated users, we can look forward to seeing many more claims of UFOs on video from perfectly sincere persons who do not have a clear idea of what can be seen in the sky.

## 2011: Infrared UFOs Over California

**"Is this a squadron of UFOs flying over California?"**, asked the *Daily Mail* of London on June 1, 2011, adhering strictly to the *First Rule of UFOlogy*: any unidentified object spotted must be presumed to be an alien spacecraft until conclusively proven otherwise. And there were other stories about this in the press. What they had in common was: somebody photographs something in the sky that he doesn't understand, and thousands of people jump to the conclusion that it's an alien spacecraft.

An anonymous photographer known only as "**KevinMC360**" took two videos of objects that are unidentified to him, using an image intensifier ("Night Vision") device, one that (unlike the human eye) is sensitive to infrared light. He posts them to **YouTube** with the suggestion that these are **UFOs** visible only in the infrared, and unknown to science. Then thousands of viewers, including "serious" journalists who should know better, think they are seeing something extraordinary, but all it means is they don't understand how these "night vision" devices work. In some cases, the anonymity of the photographer creates the strong suspicion of a hoax, but I don't think that is the case here. The video looks absolutely unmodified. This is what you see when you look into Night Vision devices.

In his video of May 26, 2011, the interesting objects are the three that fly in a triangular formation. These look very much like the triad of NOSS satellites operated by the U.S. Navy for reconnaissance. Most information about these satellites and their mission is classified for security reasons. I have seen these myself, using binoculars, and this looks very much like what I saw. Notice that the three objects keep the same position with respect to each other, except that the configuration flattens as the objects move further away. This is exactly what you would expect to see as the configuration recedes to the horizon. The single object that brightens is not too interesting, it is probably just a satellite moving into a position where its sun angle is more favorable toward the observer.

In his video of June 11, 2010, the objects behave differently. There are four of them, and they do not stay in strict formation. Probably they are birds. The problem is, however, that in low-resolution devices such as this, practically any object is simply a white dot. His argument that these UFOs must be visible only in infrared, because his wife could not see them with her naked eye, doesn't wash. The night vision device amplifies available light, infrared or otherwise, making faint objects become visible. Had his wife been using binoculars, she probably would have seen them better than he did. The video also shows the lights of a jet aircraft. Is the jet using infrared lights? Of course not.

## Sightings of UFOs

I attended the 2012 *International UFO Congress* in Fountain Hills, Arizona (near Phoenix). When I picked up my badge on the first night, there were only two things going on: a movie, and a Skywatch. The Skywatch was organized by the UFO Skywatch Club of Fountain Hills. They use very

powerful night-vision equipment to look at the night sky for moving objects. The gentleman seated at the computer has a camera pointing straight up, more or less. He moves it around a bit. It shows a lot more stars than can be seen with the naked eye, displayed on the big screen. If you want to lie back in the reclining "zero-gravity chair" and look through the night-vision binoculars for yourself it will cost $20 (although the price soon fell to $10, that being all the market would bear). They saw several airplanes and satellites. They

*Searching for UFOs using very expensive Night Vision equipment (photo by author at the International UFO Congress, Arizona, 2012)*

also saw objects that fluttered about and changed directions. Those are either owls, bats, or even possibly moths. The camera has an extremely wide-field and tremendous depth-of-field - when the operator walked up to dust something off the lens, his finger was almost in focus, as were the stars. So there was no way to tell how far away an object might be, just by looking at the display. "We're seeing a lot of UFOs tonight," one of the Skywatchers said. Pretty much whatever they might see is a UFO. Sometimes the persons demonstrating expensive night vision equipment are sales representative for the manufacturer, seeking a fat commission.

## Satellites Flash and Flare

During the first fifty years of Saucerdumb we had no shortage of objects in the sky that cause UFO sightings. Then, in the 51st year of that era we were suddenly gifted with an entirely new phenomenon guaranteed to bamboozle the casual sky watcher. A new generation of global communications technology had been developed and deployed, under the name *Iridium*. This is a series of communications satellites developed by Motorola's Satellite Communications Division to provide direct satellite-to-telephone communications, virtually anywhere on the globe.

A typical satellite in low earth orbit appears as moving "star" taking several minutes to cross the sky. Some vary in brightness if they are tumbling, typically a spent rocket booster. The brightest such satellite, the International Space Station, is usually brighter than any star when it passes over, but not as bright as Venus. The first Iridium satellites were launched on May 5, 1997, and almost immediately, amateur satellite watchers began reporting remarkable things. While the Iridium satellites are not particularly large and are normally visible only in binoculars, satellite watchers were astonished to see one or more of the Iridiums suddenly flare up to be the brightest objects in the sky for a few seconds, much brighter than Venus or Jupiter.

Mathematical analysis by satellite observers Rob Matson and by Randy John quickly yielded an answer to the mystery. The four Main Mission Antennas of the Iridium satellites, developed by Raytheon, are oriented at 90 degrees to each other. While they are not especially large (188 by 86 cm), they consist of highly-reflective aluminum flat plates, treated with silver-coated teflon for thermal control. Each being maintained at an angle of 50 degrees from the earth toward the satellite zenith, one always facing in the direction of the satellite's travel, they represent the most perfect flat reflecting surfaces ever to orbit the earth. When the angle is just right between the satellite, the observer, and the sun, sunlight reflecting off the silvered panels results in the sudden appearance of a dazzlingly-bright, slowly-moving, unexpected object that disappears in about 20 seconds - the perfect culprit to cause UFO sightings. The flare-up lasts just long enough for someone to shout "Look - up there!" It gives the crowd a few seconds of dazzling brilliance, and then fades completely from view. Flares from the Iridium satellites are visible sporadically in most locations around the globe, typically during evening or morning twilight. Now fully deployed, the Iridium program consists of 66 satellites and several spares, and the flares can be expected to continue for many years. While once expected to be widely used by businesses, the heavy phones required to communicate directly with satellites were not popular, even though they can be used in the most remote places on earth. Plus, satellite phones typically do not work indoors, or in a car.

Today the Iridium communications system is mostly used for defense and scientific applications, providing excellent communications in remote places across the globe

While we're discussing satellites that mimic UFOs, we should mention Superbird A, a dead Japanese communications satellite now adrift in the Satellite Graveyard just outside the Clarke belt of geosynchronous orbit, much more distant than the Iridium satellites. It, too, sometimes reflects the sun from both the front and back side of its huge solar panels as it tumbles out of control approximately once every 23 seconds. When the satellite is favorably placed in the sky and the sun angle with the observer is just right, for a period of about 6 minutes every evening observers can see milisecond-duration pulses of light every 11.6 seconds, looking for all the world like a strobe flash hanging in the heavens. While not nearly as bright as the Iridium reflections because of its great distance, Superbird A is the only object in or near geosynchronous orbit regularly visible to the naked eye, appearing as a puzzling stationary flash. Because of its distance from earth, Superbird A can be expected to remain in orbit for intervals measured in "eras", not "millennia". The night sky will continue to glitter with space junk long after we're all dead.

## 2010: Balloon UFOs Over Manhattan

We have been fed so many UFO promotions that every time we see some little thing in the sky, some people immediately jump to the conclusion that it's some mysterious phenomenon. The **New York Daily News** reported on October 13, 2010, "A **mysterious shiny object** floating high over Manhattan's West Side set off a flurry of reports and wild speculation Wednesday that a UFO was flying over the city. Police and the FAA said they began getting flooded with calls starting at 1:30 p.m. from people reporting a silvery object hovering high over Chelsea." "It's been hovering there for a while. I'm just kind of baffled," said Joseph Torres, 49, of Dyker Heights, Brooklyn, who spotted the object after leaving a movie. "How can it be ordinary? There is something going on."

On one of the videos posted of the object, an observer can actually be heard saying repeatedly, "they're balloons." And balloons indeed they were, according to a New York Daily News story the following day: "A Westchester elementary school believes the puzzling orbs floating over Chelsea were likely a bundle of balloons that escaped from an engagement party they held for a teacher... A parent was bringing about 40 iridescent pearl balloons to the school for language arts teacher Andrea Craparo when the wind spent a bunch away around 1 p.m."

Then some people remembered an obscure prediction by retired Air Force officer Stanley A. Fulham, who predicted a few months earlier that huge ships would be seen hovering over cities worldwide on Oct. 13, and began tweeting this sighting widely. Obviously, balloons seen in the sky must be a fleet of spaceships.

But some people, like one observer on *extraordinaryintelligence.com*, aren't buying the balloon explanation: "I was there. It was very unsettling to watch. There's no rational explanation to that spectacle in the sky. They stood together, in the same relative position to each other – you could call it a formation – and in the same spot in the sky, with very minimal movement for a long time. Small bright perfectly formed dots, against a perfectly blue sky. Surreal."

This brings us back to **The First Rule of UFOlogy:** any time you see something in the sky that you can't immediately identify, assume it's an alien spaceship until conclusively proven otherwise.

## 2010: Three Times, a "UFO" Closes an Asian Airport – What Was It?

While sightings of alleged UFOs occur pretty often, it's rare indeed for one to impact scheduled air traffic. What was the Mysterious Object that allegedly hovered over Hangzhou's Xiaoshan Airport, China's ninth-busiest, on the evening of July 7, 2010? At about 8:40 PM local time a UFO was reported by a flight crew preparing to land. As a precaution, flight controllers delayed or redirected eighteen flights.

As usually happens, conflicting and confused accounts of the UFO incident make it difficult to determine exactly what had happened. Reporters want an exciting story, and UFOlogists want to win converts, so they will typically grab onto any photo or video that is supposed to represent the object, and also report as fact practically any claim that is made. Many images soon began circulating in news stories and on the Internet, images that were supposed to show the **UFO** over **Hangzhou**, but obviously showed a different object. Most frequently seen is the impressive-looking rectangular object accompanying the **Peoples Daily** news report about "**flights diverted**," shining a beam of light down below it. At first I thought it was the reflection of an interior fluorescent light fixture seen looking out a window, but after studying the other photos in this series and reading some of the on-line discussions I agree it's probably an exposure of a few seconds showing a moving helicopter with flashing lights, shining a searchlight on the ground. In any case, since these photos had been posted to the Internet a year earlier, they obviously have nothing to do with the July 7 incident. Other photos accompanying news accounts of this incident show a rocket launch, probably that of the Russian Progress M-06M supply ship launched to the ISS on June 30.

However, it is significant that at that time, a brilliant Venus was in the evening sky, nearing its maximum brightness and setting about two and a half hours after the sun. One would think it impossible for a group of educated and seemingly rational people to mistake the brilliant planet Venus for a UFO – but experience shows otherwise. "No single object has been misinterpreted as a 'flying saucer' more often than the planet Venus. The study of these mistakes proves quite instructive, for it shows beyond all possible dispute the limitations of sensory perception and the weakness of accounts relating shapes and motions of point sources or objects with small apparent diameters," wrote the well-known pro-UFOlogist Jacques Vallee in his 1966 book *Challenge to Science*.

Officials said that the object did not turn up on radar. Ruan Zhouchang, spokesperson for the Xiaoshan Airport in Hangzhou said through an interpreter, "There was an unknown object seen in the skies over the airport. So according to our regulations we had to close the airspace. Aircraft movements were suspended from 8:45 PM to 9:41 pm. " However, the photo accompanying that news report, supposedly showing the object, appeared to be that of a high-altitude contrail .

A 2001 incident occurring at the Barnaul Airport in southern Siberia sounds very much like the incident at the Xiaoshan Airport. The Agence-France Press reported from Moscow on Jan. 27, 2001: "An airport in southern Siberia was shut down for an hour and a half on Friday when an unidentified flying object (UFO) was detected hovering above its runway, the Interfax news agency reported. The crew of an Il-76 cargo aircraft of the company 'Altaï' refused to take off, claiming they saw a luminescent object hovering above the runway of the [sic] Siberia's Barnaul airport, local aviation company director Ivan Komarov was quoted as saying." As in Hangzhou, nothing appeared on the radar. UFO investigator Eric Maillot from the Cer*cle Zetetique*, a French rationalist organization, looked into this case. He discovered that the time and direction of the UFO matched the position of Venus. And since the crew reported seeing just one bright object, not two, Maillot concluded that the "UFO" reported by the crew **was** Venus, and not some separate object .

An official investigation by Chinese aviation experts was launched into the incident at Xiaoshan Airport. They concluded that the object probably was "an airplane on descent reflecting light," one which apparently kept descending for an hour or more. They said it might also have been an "unexpected flying vehicle," whatever that is. Imagine how embarrassing it would be to have to conclude, "Air Traffic Controllers at one of our major Chinese airports mistook Venus for an unknown hovering aircraft."

In yet a third incident where a "UFO" has closed an airport, Venus is again a prime suspect. The incident occurred on September 11, 2010 in Baotou, Inner Mongolia, but wasn't reported until several weeks later. As reported in *The Telegraph* (London, October 5, 2010), "An airport in Baotou, Inner Mongolia, was forced to shut to prevent passenger jets crashing into a UFO, according to reports. Starting around 8 PM, three flights to Baotou from Shanghai and Beijing were reportedly forced to circle the airport until the UFO disappeared. Two other flights were diverted away from Baotou and to the nearby cities of Ordos and Taiyuan. The airport was shut for around an hour "to guarantee safety" according to a spokesman."

On September 11, Venus was setting about 1 hour 20 minutes after the sun, from the latitude of Baotou, Inner Mongolia. Notice that the airport was closed for "around an hour" until the object disappeared.

We cannot say with certainty that the "UFO" was Venus, because we do not have information about the object's apparent elevation and direction. But past experience creates the suspicion. At first it might seem impossible that educated and sane people would mistake the bright planet Venus, then near maximum brilliance, for a hovering UFO, but it has happened over and over again, all over the world.

# 2008: Mysterious Spiral UFO over Norway

"Mystery as spiral blue light display hovers above Norway" proclaimed the *Drudge Report*. "A MYSTERIOUS giant spiral of light that dominated the sky over Norway this morning has stunned experts — who believe the space spectacle is an entirely new astral phenomenon," exclaimed the British tabloid newspaper *The Sun*. And dozens of other news publications and websites ran similarly breathless stories. All across Norway on the morning of December 9, people heading off to work saw a bright "star" rising in the sky, spiraling out an amazing pinwheel of white light. A blue tail seemed to point back to earth. The well-known British UFOlogist Nick Pope told the Sun, "A meteor or a fireball would simply travel in a straight line but for something to spiral in this way appears to go against the laws of physics...It's ironic that something like this should happen the very week after the MoD [British *Ministry of Defence*] terminated its UFO project. It just goes to show how wrong that decision was." For about 24 hours, the media extolled the great mystery.

While the mystery-mongers were busy spreading their uninformed speculations, serious observers immediately suspected it was a Russian rocket launch. At first the Russians denied any launch, but the denials had no plausibility, since Russia had just posted a warning for ships to stay out of that area because of a scheduled missile launch. Also, similar launch spirals had been observed before, when in 2007 a Russian rocket spiraled out of control. About a day later, the Russian Defense Ministry acknowledged the failed test launch of its troubled new Bulava series ICBM.

The incident occurred sometime after 8:00 AM, when the sky was still dark. Photos of the incident show twilight beginning (in Norway during the winter, the sun doesn't rise until after 9:00). As the rocket rose higher above the ground, it entered sunlight while ground observers were still in darkness. When the rocket began venting propellant while spinning, the result was a spectacularly-bright sunlit spiral unwinding, seen against a dark sky. I can truly say that this is the most spectacular "UFO" that has ever been captured on photo or video, and our knowledge of what it is does not diminish its awe.

## 2010: Mystery Missiles – or Aircraft Contrails?

Residents of the tiny town of **Harbour Mille,** along the southern coast of Newfoundland in Canada's Maritime Provinces, were reporting, and photographing, what were called "**missiles**" flying just off the coast. It began about 5PM on January 25, 2010 when Darlene Stewart went outside to photograph the sunset, and saw "a long, thin glimmering object in the sky that appeared as if it came out of water." She called her friend over, who later stated, "I saw a humungous bullet, silver-grey in colour and it had flames coming out of the bottom and a trail of smoke." The object remained in sight for about fifteen minutes (far too long to be any kind of missile).

Some attributed the sighting to small rockets launched by amateurs, who sometimes launch from that area. Liberal Senator George Baker found a politically acceptable explanation for the matter: blame France, whose nearby island of St-Pierre-Miquelon was in the direction of the sightings. "Knowing that France has territory within our 200 mile zone in Canada, they should at least ask the French, 'Look, are you launching these missiles?' Because if they are, (and) everybody is denying knowledge of it, then the laws have been broken." But the French replied that they had not launched any missile anywhere on that day. So what kind of missile was it?

None at all, said Finnish UFO researcher **Bjorn Borg**. He calls it "the December Phenomenon" because "Every year this comes up in the news." What people are actually seeing, he says, are jetliner **contrails** catching winter sunlight, an optical illusion created when the angle between the setting sun and the jetliner is just right. "The sun is shining on the (condensation) trail. In winter time, the colour of the trail will show up (in) very strong yellow or even red. It looks like fire." Chris Stevenson, the local president of the Royal Astronomical Society in Newfoundland, who said "I've seen this several times." He explained that transatlantic aircraft reflect sunlight far overhead after daylight has faded on the ground, appearing a deep reddish color. Besides, "A rocket plume would be thick all the way back to where the rocket was launched, " he explained.

On Nov. 9, 2010 the news media were again filled with reports of a "mystery missile," this time allegedly fired off the California coast near

## Sightings of UFOs

Los Angeles. The *Drudge Report* headline screamed "MYSTERY MISSILE FIRED OFF CA COAST; PENTAGON 'NO CLUE'. CBS News in Los Angeles was reporting, "A **mysterious missile launch off the southern California coast** was caught by CBS affiliate **KCBS**'s cameras Monday night, and officials are staying tight-lipped over the nature of the projectile. CBS station KFMB put in calls to the Navy and Air Force Monday night about the striking launch off the coast of Los Angeles, which was easily visible from the coast, but the military has said nothing about the launch." It certainly looks like a missile launch! But is it? Not only did the Pentagon deny all knowledge of any possible launch, but if a foreign country were to fire a missile so close to our shores, it would be an act of war. By saying that officials "are staying tight-lipped over the nature of the projectile," CBS news was implying that A), the object must have been a "projectile" or rocket, and B) that military officials knew what it was, but refused to tell the truth to the public.

Again, the object seems to have been simply an aircraft contrail. Tricks of perspective made it look like a missile flying away from you, when in fact it was an aircraft flying toward you This is not the first time such a thing has happened in California. On Dec. 31, 2009, much the same excitement occurred off the coast just south of Los Angeles in Orange County. A similar article was published in the **Orange County Register** about the "**mystery launch**" of December 31, **2009**. Remember, this is not the incident of November 8, 2010, but one occurring about ten months earlier. Yet the explanation is the same. Because of the viewing angle and the exceptionally clear skies, a contrail of an aircraft headed directly toward the observer looked like a missile being launched. Mick West's blog entry on his **ContrailsScience**.com shows the **2009 Orange County** contrails and the one in Los Angeles the following year were quite the same. That same article has several very interesting incidents of contrail hysteria in the U.S., going back as far as 1950.

The appearance of the contrail depends on effects of perspective. The aircraft's path must be directly toward, or away from, the observer. Second, even though the contrail is five miles above the ground, as it recedes into the distance it appears to touch the ground, because of the curvature of the earth. As shown by the daytime photo of the vertical contrail on ContrailsScience.com, we know that the aircraft that made it was not flying straight up like a rocket, but when seen directly straight on, that is what it looks like. And for viewers a few miles away, getting a different perspective, all they see is an ordinary-looking slanted contrail.

And the award for Best Pompous Pontification by an Uninformed Fool goes to former U.S. Ambassador to NATO Robert Ellsworth, also a former Deputy Secretary of Defense, who was happy to appear on TV and speculate that, because President Obama is in Asia, "It could be a test-firing of an intercontinental ballistic missile from a submarine ... to demonstrate, mainly to Asia, that we can do that." As if anyone in Asia seriously doubted that the U.S. could launch missiles from a submarine! Ellsworth added that ICBM testing was carried out in the Atlantic to demonstrate America's power to the Soviets during the Cold War, but he says he doesn't believe an ICBM has been tested by the U.S. over the Pacific. I have two words for him: Vandenberg and **Kwajalein**. Such **ICBM tests** over the Pacific occur on a regular basis. So much for seeking "informed comments" from a former Deputy Secretary of defense.

Although the so-called "mystery missile" photographed off Los Angeles on was quickly identified as an aircraft contrail by responsible investigators, apparently it has morphed into a "mystery" that's too good to give up. No matter that earlier instances of "contrails scares" have been identified and explained not only in California, but elsewhere in the world, going back as far as 1950, no matter that the exact flight responsible for the contrails has apparently been identified – **contrailsscience.com** demonstrated that it was **UPS flight 902** from Hawaii to the **Ontario Airport** in California. No matter that a nearly-identical contrail from that same daily flight was recorded exactly 24 hours later on a webcam in Newport Beach - conspiracy theorists simply will not let the "mystery" die. This story has apparently reached critical mass, so like Roswell, the JFK assassination, etc., *it no longer matters what the facts are.* The facts are out in plain view for all to see. But conspiracy theorists reject facts that are public and demonstrable, and substitute their own.

For example, an article in the conspiracy-oriented publication the *Los Angeles Times* (November 10, 2010) proclaims "Puzzling lack of answers to 'Mystery Missile.' " In a piece worthy of Erich von Daniken, the Times reports "Military and aviation authorities deny any knowledge of a scheduled launch off the coast of L.A. The Pentagon says only that it is looking into a report of an 'unexplained contrail' left by an aircraft." Well, if the Pentagon says only that it might be a jet contrail, then they're obviously covering something up. "Some aerospace experts who reviewed the footage said the size of the plume hinted that it was a government operation." (More uninformed "experts" blubbering nonsense.) "It can't belong to anyone but the military," said Marco Caceres, an analyst with Teal Group Corporation, a Fairfax, Va.-based aerospace research firm. The appearance of such a massive rocket contrail near military bases that are known for regularly testing missiles is unlikely to be a coincidence, Caceres said." A hilarious piece in the *Los Angeles Times* (November 9, 2010) proclaimed "L.A. 'mystery missile' may have been errant launch, experts say," or "Honey, I launched the kids."

Again, "experts" who have no actual information on the subject are perfectly happy to talk rubbish. Before speculating on "where" the alleged missile came from, one first needs to determine "whether" a missile was, in fact, filmed. Let's see: it was moving too slowly to be a missile, there was no bright glow from a rocket engine (if you've ever seen an actual rocket launch, you won't forget that sight), nothing unexpected turned up on any radar, and nobody saw or photographed anything except the guy in the traffic helicopter. Sounds to me like a North Korean missile.

Meanwhile, the comments on various internet Blogs, forums, etc. weigh in heavily in favor of the "they won't tell us the truth" persuasion. The favorite explanations are an "accidental launch" from some U.S. ship, or else a Chinese or North Korean submarine.

And in the classic manner of UFO-contagion, Contrail Hysteria spread to New York City. A CBS helicopter there filmed on November 10, 2010 a "bizarre, glowing red-hot streak in the sky — right at sunset Wednesday — moving briskly behind the Manhattan skyline." And that one isn't even very impressive, as it's horizontal not vertical, but hey, it's glowing a fiery red color at sunset! It seems that "Mystery Missiles" are being sighted everywhere, and conspiracy tales about them abound. A **Mystery Missile** was soon reported to be attacking **Queens, New York**. It seems that enemy submarines are turning up everywhere. But this plume was *west* of New York City - meaning that enemy submarines have now figured out how to operate on land!!

Conspiracy theories continued to proliferate. **Doug Richardson**, the editor of Jane's Missiles and Rockets (who should know better), told the *Times* of London that the **Mystery Missile** might have been a Standard interceptor, the anti-missile weapon which is fitted to the US Navy's Aegis guided-missile cruisers as part of the American missile defense program. "It's a solid propellant missile, you can tell from the efflux [smoke] but they're not showing enough of the tape to show whether it's staging [jettisoning its sections]."

Sightings of what were called "ghost rockets" in Europe, mostly in Sweden, in the aftermath World War II, is well known to students of UFOlogy, if not to the general public. They were "ghost rockets" in the sense that people reported seeing objects they described as "rockets" or "missiles," yet no evidence of the actual existence of such objects has ever turned up. (See the **Wikipedia** article on **Ghost Rockets**). Some of the objects were reported to crash into lakes (why was it never a farmer's field?), but despite numerous searches no rocket parts were ever recovered from any lake, or anywhere else.

While I was reflecting on the current epidemic of sightings of "mystery missiles", not only in California but elsewhere, it occurred to me that people were actually reporting "ghost rockets" today. After all, a "mystery missile" is no different from a "ghost rocket." Indeed, it seems likely that some future book about UFOs will contain a sentence along the lines of, "Sightings of 'ghost rockets' began occurring in the U.S. and Canada around 2010." Thus far, all of these North American Ghost Rockets appear to be attributable to high-altitude contrails from jet aircraft.

Were there any high-altitude aircraft capable of creating contrails flying around Scandinavia in 1946? Large planes flying at high altitudes, such as the B-17 or B-29, will typically produce contrails when meteorological conditions are favorable. There is no need for them to be jets. A dispatch from the **U.S. Naval Attaché in Stockholm**, dated **August 16, 1946** and formerly classified "Top Secret," talks of "civilian observers reporting jet fighters, contrails and meteors as rockets." The U.S. used the high-altitude B-29 bomber for reconnaissance in the arctic following World War II, as well as in Europe.

Some of the "ghost rockets" were surely meteors, especially those seen at night, described as fast-moving and only seen for a few seconds. Still others were probably astronomical objects, described as bright lights hovering in the night sky. People tend to scrutinize the heavens more than usual when they have heard that unusual objects are zipping about. But other "ghost rockets" were described as moving much more slowly, and flying horizontally. These sound much more like contrails. The single "classic" photo of a Swedish "Ghost Rocket," seen in the Wikipedia article, is usually attributed to a meteor, but looks very much like a high-altitude contrail. Notice the cirrus clouds above it. "Contrails are a form of cirrus cloud," ContrailsScience.com reminds us. Indeed, the meteorological conditions that produce contrails are the same as those producing cirrus clouds. If cirrus clouds cannot be produced, then neither can contrails.

It has often been suggested that the Swedish "ghost rockets" of 1946, reports of which were carried worldwide, played a role in creating the "flying saucer" excitement that broke out over Kenneth Arnold's sighting the following year. And thus, in creating the entire UFO scenario. So, what I'm suggesting is that the "ghost rockets" excitement of the present seems to be a replay of the earlier Swedish excitement. We know from present experience that jet contrails can fool even some very sophisticated people into believing that they are seeing rockets or missiles, and this in a time when contrails are already a very familiar sight. We recall that ContrailScience.com listed incidents of contrail hysteria in the U.S., going back as far as 1950. That's getting awfully close to 1946.

Thus it seems quite likely that the main stimulus behind the "ghost rockets" of 1946 was the presence of contrails in the sky, in a time when that phenomenon was new and not at all familiar. World War II had ended just the previous year, and it was known that the U.S. and the USSR were both frantically pursuing missile development, using captured German rocket scientists. Everybody knew that the Cold War could turn into a "hot war" at any moment, and on several occasions it nearly did. Sweden and the rest of Scandinavia were literally in the middle of it, between the allies US/UK/France, and their adversary, the Soviet Union. It is no wonder that, in such an atmosphere, people in Sweden, seeing the unfamiliar new phenomenon of high-altitude contrails, sometimes perceived them as menacing rockets launched by one great power or another. Those in the U.S. and Canada today, however, who make up conspiracy stories in the same vein, even in situations where "ghost rockets" would have to have been fired on land, do not have the extenuating circumstances that the Swedes had more than sixty years earlier.

# UFO Sightings Everywhere

UFOs that seem to be genuinely extraterrestrial have been spotted in Missouri. In June, 2012 KCTV5 in Kansas City broadcast a news story about sightings of "strange lights in the sky" seen hovering over the neighborhood of Blue Springs. Some of these lights have seen "for weeks," "vibrating lights, red, green, and blue." One UFO investigator came out to see them, and proclaimed "I'm 90% certain that we're looking at Vega in this instance." But she wasn't certain because she had been told by a colleague that Vega is bluish, and this object had sparkles of red and green color. Had she taken an astronomy class, she would have known that the earth's atmosphere causes stars to twinkle, and breaks down starlight into flashes of different colors. Reporter Dave Jordan said that he contacted the Blue Springs Police, the FAA, and NORAD, and none of them had any information on these lights. NORAD had, however, received one other UFO report, and "is still working to determine whether that report came out of Missouri." Let us hope that NORAD completes this difficult investigation quickly, and reports its finding.

Missing from that contact list is "astronomer." Any astronomy professor, or even an advanced amateur, could have immediately identified which stars, and which planets, were being spotted as "UFOs."

## Aircraft Captured by UFO?

A famous allegedly "unexplained" UFO case is the 1978 disappearance of Fredrick Valentich, a 20 year old pilot in Australia. On October 21 1978 he was piloting a Cessna 182L light aircraft over Bass Strait in Australia. He intended to land at King Island and return to Moorabbin Airport. However, he never made it to King Island, 127 miles away. The final exchanges between Valentich (DSJ) and air traffic control are as follows: (from Wikipedia)

> 19:06:14 DSJ [Valentich]: Melbourne, this is Delta Sierra Juliet. Is there any known traffic below five thousand?

> FS [Flight Services; Robey]: Delta Sierra Juliet, no known traffic.

> DSJ: Delta Sierra Juliet, I am, seems to be a large aircraft below five thousand.

# Sightings of UFOs

19:06:44 FS: Delta Sierra Juliet, What type of aircraft is it?

DSJ: Delta Sierra Juliet, I cannot affirm, it is four bright, and it seems to me like landing lights.

FS: Delta Sierra Juliet.

19:07:31 DSJ: Melbourne, this is Delta Sierra Juliet, the aircraft has just passed over me at least a thousand feet above.

FS: Delta Sierra Juliet, roger, and it is a large aircraft, confirmed?

DSJ: Er-unknown, due to the speed it's travelling, is there any air force aircraft in the vicinity?

FS: Delta Sierra Juliet, no known aircraft in the vicinity.

19:08:18 DSJ: Melbourne, it's approaching now from due east towards me.

FS: Delta Sierra Juliet.

19:08:41 DSJ: (open microphone for two seconds.)

19:08:48 DSJ: Delta Sierra Juliet, it seems to me that he's playing some sort of game, he's flying over me two, three times at speeds I could not identify.

FS: Delta Sierra Juliet, roger, what is your actual level?

DSJ: My level is four and a half thousand, four five zero zero.

FS: Delta Sierra Juliet and you confirm you cannot identify the aircraft?

DSJ: Affirmative.

FS: Delta Sierra Juliet, roger, stand by.

19:09:27 DSJ: Melbourne, Delta Sierra Juliet, it's not an aircraft it is (open microphone for two seconds).

19:09:42 FS: Delta Sierra Juliet, can you describe the - er - aircraft?

DSJ: Delta Sierra Juliet, as it's flying past it's a long shape (open microphone for three seconds) cannot identify more than it has such speed (open microphone for three seconds). It's before me right now Melbourne.

19:10 FS: Delta Sierra Juliet, roger and how large would the - er - object be?

19:10:19 DSJ: Delta Sierra Juliet, Melbourne, it seems like it's chasing me.[21] What I'm doing right now is orbiting and the thing is just orbiting on top of me also. It's got a green light and sort of metallic like, it's all shiny on the outside.

FS: Delta Sierra Juliet

19:10:46 DSJ: Delta Sierra Juliet (open microphone for three seconds) It's just vanished.

FS: Delta Sierra Juliet.

19:11:00 DSJ: Melbourne, would you know what kind of aircraft I've got? Is it a military aircraft?

FS: Delta Sierra Juliet, Confirm the - er ~ aircraft just vanished.

DSJ: Say again.

FS: Delta Sierra Juliet, is the aircraft still with you?

DSJ: Delta Sierra Juliet; it's (open microphone for two seconds) now approaching from the south-west.

FS: Delta Sierra Juliet

19:11:50 DSJ: Delta Sierra Juliet, the engine is rough-idling. I've got it set at twenty three twenty-four and the thing is (coughing).

FS: Delta Sierra Juliet, roger, what are your intentions?

# Sightings of UFOs

DSJ: My intentions are - ah - to go to King Island - ah - Melbourne. That strange aircraft is hovering on top of me again (open microphone for two seconds). It is hovering and (open microphone for one second) it's not an aircraft.

FS: Delta Sierra Juliet.

19:12:28 DSJ: Delta Sierra Juliet. Melbourne (open microphone for seventeen seconds).

It was Valentich's first and only night flight over water. And neither Valentich nor his aircraft was ever seen, or heard from, again.

At last we have some new information on this puzzling case: "Adelaide researcher Keith Basterfield has been following the case since the disappearance in 1978, but had been told by the Government in 2004 the official file had been lost or destroyed. He "found" it when searching through an online National Archives index on an unrelated topic. The file has since been digitised and uploaded on the archive's website" (http://tinyurl.com/butdpmv).

So we have skeptic Keith Basterfield to thank for the recent government "document dump" that makes this new information available. Basterfield explains that the newly released files reveal that "parts of aircraft wreckage with partial serial numbers were found in Bass Strait five years after the disappearance." Also, one pilot searching at the right time and place saw debris that appeared to be from a Cessna, but before he could get a good fix on its position it apparently sank. This makes it extremely likely that Valentich's aircraft simply crashed into the water in the darkness.

A number of reports of a fast moving brilliant white light were received from various parts of the country. Mt Stromlo observatory advised that the night of the 21st was the peak of the meteorite stream with 10-15 sightings per hour achieved.

The question of why Valentich took this somewhat risky night flight is a separate matter. According to Wikipedia,

His stated intention was to fly to King Island in Bass Strait via Cape Otway, to pick up passengers, and return to Moorabbin. However, he had told his family, girlfriend and acquaintances that he intended to pick up crayfish. During the accident investigations it was learned there were no passengers waiting to be picked up at King Island, he had not ordered crayfish and could not have done so because crayfish were not available anyway.

So clearly Valentich was being evasive about something. Also, it turns out that Valentich was a UFO True Believer, and hence probably inclined to assume anything as a "UFO" that he could not immediately identify. He actually worried about what to do if a UFO attacked him!

Various air safety organizations including the Aircraft Owner and Pilots Association have a pilot education lesson titled **178 Seconds to Live**. The title comes from a study that found (using flight simulators, of course) that when non-instrument rated pilots are placed in situations where they cannot see the ground or the horizon, nearly all of them got into a fatal spiral and crashed, with an average time-to-crash of 178 seconds. Watching that video made my blood run cold, but it's absolutely realistic.

This is exactly the situation Valentich found himself in when darkness fell on that moonless night (as did JFK Jr in 1999). Assuming that he became disoriented and thought that Venus, or perhaps a meteor, was a UFO – he says it was "orbiting" him – we would expect him to crash in about 178 seconds. He actually survived 374 seconds from the time of his first UFO report until crashing. Valentich had a "Class Four Instrument Rating," but we know he was not watching his instruments; his eyes were fixed on the "UFO" he was describing. We also learn from Wikipedia that Valentich

> had twice applied to enlist in the Royal Australian Air Force but was rejected because of inadequate educational qualifications. He was a member of the Air Training Corps, determined to have a career in aviation. His student pilot licence was issued 24 February 1977 and his private pilot licence the following September. Valentich was studying part-time to become a commercial pilot but had a poor achievement record, having twice failed all five commercial licence examination subjects, and as recent as the previous month had failed three more commercial licence subjects. He had been involved in flying incidents, straying into a controlled zone in Sydney (for which he received a warning) and twice deliberately flying into cloud (for which prosecution was being considered).

I would never knowingly get into an aircraft with a pilot like that – and especially not for a night flight over water! We can be quite sure of what happened to the incautious Valentich, even if we cannot say why he made that fatal flight.

## The RB-47 Radar Case: UFOlogy's Best Evidence?

In the early morning hours of July 17, 1957, a U.S. Air Force crew aboard an RB-47, a plane loaded with the most sophisticated state-of-the-art surveillance and electronic countermeasures gear, reportedly encountered and was followed across several southern states by one or more UFOs, seen visually as well as on radar. Some UFOlogists consider this the best UFO evidence of all time. It was investigated as Case 5 of the University of Colorado's Air-Force-sponsored Condon Report, which cited the absence of an official report supposed to have been written on the incident, and concluded "Evaluation of the experience must, therefore, rest entirely on the recollection of crew members ten years after the event. These descriptions are not adequate to allow identification of the phenomenon encountered."

The UFO Casebook says, "An Air Force RB-47, equipped with electronic countermeasures (ECM) gear and manned by six officers, was followed by an unidentified object for a distance of well over 700 miles, and for a time period of 1.5 hr., as it flew from Mississippi, through Louisiana and Texas and into Oklahoma.

"The object was, at various times, seen visually by the cockpit crew as an intensely luminous light, followed by ground-radar and detected on ECM monitoring gear aboard the RB-47.

"Of special interest in this case are several instances of simultaneous appearances and disappearances on all three of those physically distinct "channels," and rapidity of maneuvers beyond the prior experience of the aircrew.

There is a brand new in-depth investigation of this case by Tim Printy, published in his WebZine *SunLite*, January/February, 2012. It is one of the most complex cases in all UFOlogy. I cannot possibly give more than a brief summary here; Printy's analysis runs over thirty pages, and is enormously significant in the history of this major case, and thus in the ongoing debate over the reality of UFOs.

Printy begins by making what is, to me, a crucial observation: "It does seem rather odd that the UFO would decide to use an S-band radar signal to track or test an Air Force RB-47. It is this clue that seems to have been glossed over/down played by those presenting this case as the best evidence." In other words, the UFO seems to have been sending out (but only in this case) S-band radar signals with exactly the same characteristics as those used by the U.S. Air Force at that time. How strange is that if the UFO is sending out exactly the same kind of radar signals we do? So isn't it likely, then, that the source of the signals was not an extraterrestrial craft, but instead a misidentified terrestrial one?

Printy notes how the atmospheric physicist and Ufologist Dr. James E. McDonald (1920-1971) interviewed the RB-47 crew and wrote a paper on this case. "McDonald's stamp of approval had immediately made this case a "classic."" The famous UFO skeptic Philip J. Klass, "took on the case in 1971 and wrote a rather extensive study on the incident. Klass suggested that it was equipment malfunction, a bright fireball, an airliner, and reception of ground radar signals that made the event appear mysterious to the air crew." This analysis can be found in Chapters 19 and 20 of Klass' 1974 book *UFOs Explained*.

In the 1990s, UFOlogist Brad Sparks, who styles himself as "the "RB-47 expert" in his email address, re-evaluated the case. Sparks concluded that Klass had erred, most especially in asserting that the RB-47 had, because of an equipment malfunction, erroneously picked up signals from the radar station at Kessler Air Force base in Biloxi, Mississippi, which was a training facility for radar operators and repairmen. That radar, said Sparks, was not operating at the time of the incident! Sparks wrote, "Since it was a nine-month course it was apparently run during the normal academic term from September to June approximately. In other words, there would not have been a class in session to operate the CPS-6B even in the daytime, let alone nighttime, in the midst of summer vacation, on July 17, when the RB-47 incident took place."

Sparks simply assumed that, since this was not during the "academic year," the training radar would have been turned off! But Printy looked very closely into the matter of whether or not the Kessler radar was operational at the time in question. He found (as he, a retired Navy submariner, already surely knew), that military training schedules bear no resemblance at all to those of colleges! In fact, Printy found that the Kessler facility had been operating at maximum training capacity in July 1957, and that preventive and corrective maintenance was usually done on the midnight shift. In order to get the radar set ready for the 0600 classes, the crew had to turn on the radar to get everything configured.

Printy found that while Klass' analysis contains some errors, his overall conclusions still stand. If you have any interest in the RB-47 controversy, or if you want to read one of the very finest research papers ever published about any UFO case, then I encourage you to read Printy's paper, and come to your own conclusion about this important and controversial UFO case.

# 3. UFO Photos and Videos: Is Seeing Believing?

Photos of alleged UFOs have played a significant role since the early years of UFOlogy. In the 1950s, the famous cont actee George Adamski produced a number of photos of what he said were the space ships of his friends from Venus, some of which were supposedly taken at close range and others using his telescope. However, Adamski's photos never looked convincing, and few outside his circle of followers doubted that they had been fabricated using quite ordinary objects. It has often been said that Adamski's most famous "Venusian Scout Ship" photo was made from a chicken brooder. However, researcher Joel Carpenter (1959-2014) wrote a paper in 2012 suggesting that it was actually made using the top of a propane lantern, widely sold by Sears Roebuck and other retailers since the 1930s. These lanterns were commonly found in campsites, such as the one on Palomar Mountain where Adamski served food to the campers and tourists.

Certain "classic" UFO photos continue to have a wide following today among "Science Fiction" UFOlogists who defend them energetically. The Brazilian **Trindade Island UFO photos** of 1958 have been widely touted even though the man who took them was a **specialist in trick photography**. The **Lucci brothers' photos** from Pennsylvania in 1965, famous for being used in many **UFO** books and magazines, have recently been confessed by one of the brothers to be **hoaxes**.

There is still considerable interest in UFO photos and videos today, although today a UFO photo usually means "pixels," not "film." Many of the recent "unidentified" objects in them appear as simply dots, blips, or lights. Given the near-ubiquitous availability today of cell-phone and digital cameras, many of which are capable of producing videos, it is most curious that we do not have clear, close-up authentic photos and videos of the many reported close encounters and abductions. We do get, however, plenty of photos of blips and dots that could be practically anything. Some researchers call these **Blurfos** (blurry UFOs).

With the proliferation of software such a Photoshop, Gimp, etc. for altering photos and videos or even creating them from nothing, a photo or video cannot simply "stand by itself" as evidence of anything. For a photo or video to be convincing, we must know a great deal about its origins, the photographer, the location, etc. A number of really clever digital photo and video UFO hoaxes have been created in recent years, but typically they are submitted anonymously via the Internet – typically, posted to *YouTube.com* - because the story of their origin would not withstand scrutiny. The British UFO researcher who goes by the name **Isaac Koi** has compiled a large on-line collection and analysis of supposed **alien photos** and **UFO videos**. He not only explains how they were made, but in many cases identifies the hoaxer responsible for each.

In recent years there have been significant new findings about certain "classic" vintage UFO photographic cases. These developments are discussed below.

## McMinnville, Oregon, 1950: farmer Trent slings up a truck Mirror?

On May 11, 1950, farmer Paul Trent of McMinnville, Oregon snapped two photos of an object that he claimed was a flying saucer (the term "UFO" hadn't been invented yet). There are inconsistencies in Mrs. Trent's accounts of where her husband was when the object was first spotted, and who went inside to get the camera. They did not immediately tell anyone about the photos, or rush them off to be developed. Instead, the film containing the invaluable flying saucer photos was left in the camera until Mother's Day, so that a few unexposed frames would not be wasted. More information on the **Trent** photos is on my **debunker** website, as well as on my **BadUFOs** Blog.

The Trent photos from Oregon in 1950 tentatively passed muster with the famously skeptical Condon Report, whose analysis suggested that while the object might be large and distant, The Trent UFO could also be a small model, slung from overhead telephone wires. However, that analysis depends on certain assumptions. If the photos were fabricated using a truck mirror with a reflective surface (as now seems likely), the assumptions are incorrect.

## Bad UFOs

In 2004, researcher **Joel Carpenter** created a website on the McMinnville photos, making a very good case that the object was directly beneath the overhead wires, and close to the camera. He suggests that the object was a mirror from an old truck. I have restored Joel Carpenter's original **McMinnville** photos website (fixing only the links, also adding his Adamski paper), and placed it on the **Internet Archive**.

One of Carpenter's findings is that Trent's camera was surprisingly close to the ground when the photos were taken. For some bizarre reason, Trent did not stand up but instead crouched down to photograph his UFO. Carpenter explains,

> Instead of moving toward the object and shooting the photos from eye level in the unobstructed front yard, he shot the two photos up, from a very low level, from the back yard. For reasons explained above, it seems likely that he actually used the viewfinder on the body of the camera while kneeling. The overall geometry of the positions and the attributes of the camera suggest that he was attempting to frame a nearby object in such a way as to maximize the amount of sky around it and enhance its apparent altitude.

The two **Trent** photos as a stereo pair, created by "**Blue Shift**" on **Above Top Secret.** Cross your eyes to make the two images of the "saucer" merge, and you will see that the object is much closer than distant objects in the photos.

In other words, Trent walked away from where the UFO was supposed to be, and instead walked toward where the presumed model was hanging from the wires, and crouched down close to the ground to make his "UFO" appear distant.

## 1965: Rex Heflin, a Classic UFO Photo, now in 3-D!

In Santa Ana, California on August 3, 1965, highway worker Rex Heflin (died 2005) got three photos of a supposed UFO out the window of his truck, using his Polaroid instant camera. This object supposedly flew right over the Marine Corps El Toro Air Station, plus the Santa Ana Freeway (Interstate 5) in broad daylight, but no one else saw it. In these photos, distant objects are hazy because of the Los Angeles smog, while the UFO is not, probably because it is tiny, and very close to the camera. Unfortunately, the original prints cannot be investigated, because Heflin claims that they were confiscated by an investigator who came to his house, presenting ID supposedly from NORAD. At least Heflin did not claim that the Men In Black came for his prints, or that the dog ate them. It was NORAD. So all we had left were copies made from the originals. Nonetheless, this series of photos has long been touted as a "classic" by NICAP and by many prominent UFOlogists.

Skeptics have argued that Heflin's UFO appears to be a tiny model, just a few inches in size, hanging from something like a fishing pole propped up over the cab of his truck. The investigator for the Condon Report, William K. Hartmann, convincingly replicated these photos in exactly that manner. However only recently did a still-anonymous person, using the alias Enkidu, make an extremely important finding. In a 2006 discussion thread on the conspiracy-oriented website **Above Top Secret**, **Enkidu** argues that **Heflin** unintentionally created a **3-D photo** of his **UFO**. Assuming that the UFO was attached in some way to the truck, by moving the camera a few inches between the exposures, Heflin has produced a near-perfect stereo pair. The photos below are reversed by Enkidu to allow easier viewing of the 3-D effect without a stereo viewer by simply crossing one's eyes. And when you do that, the UFO is seen to be *tiny*. It's clearly farther away than the truck's mirror, but much closer than the roadside vegetation, or the distant trees. Responding to criticism, Enkidu writes, "Yes, it's possible that the UFO moved between the time the first photo was taken and the second. But it would have to move exactly horizontal to the way the camera moved, because there's no apparent difference in the size of the top part of the ship. It could only tilt forward. It didn't go up or down, and it didn't get nearer or closer. The odds of that happening are pretty slim." Great work, Enkidu!

Cross your eyes and align these words to see the 3-D.    Cross your eyes and align these words to see the 3-D.

"Enkidu" made this stereo pair of Heflin's two photos. The "UFO" is small, and close to the camera.

## 1989: UFOs Invade Belgium

For more than twenty years, UFO believers have been citing the 1989-1990 wave of UFO sightings in Belgium as an unexplained mystery. For a period of several months, people in Belgium were reporting sightings of a triangular-shaped craft. It was one of the major chapters in Leslie Kean's 2010 best-selling book, *UFOs: Generals, Pilots, and Government Officials Go on the Record*. Even skeptic Michael Shermer's review of Kean's book suggested that the Belgian sightings represent a "residue of anomalies" (*Scientific American*, March 28, 2011).

One big problem with the Belgian wave has always been the lack of photos or movies showing the object, despite hundreds of claimed sightings. Indeed, Kean seeks to dismiss the lack of evidence by noting that "twenty years ago, cell phones and relatively inexpensive, consumer-level digital and video cameras were not yet in use" (true, but film cameras were plentiful and widespread). Indeed, only one photo claiming to show this supposed 'triangular craft' has ever been seen. It was said to have been taken in Petit Rechain, Belgium in April 1990 by a twenty-year-old man known only as "Patrick," although it was not released until four months later. The Belgian UFO investigative group SOBEPS investigated the photo and found it to be authentic. So did many other "experts". Kean writes,

> A team under the direction of Professor Marc Acheroy discovered that a triangular shape became visible when overexposing the slide. After that, the original color slide was further analyzed by Frangois Louange, specialist in satellite imagery with the French national space research center, CNES; Dr. Richard Haines, former senior scientist with NASA; and finally Professor Andre Marion, doctor in nuclear physics and professor at the University of Paris-Sud and also with CNES. (p. 30)

UFO skeptics have long supplied reasons why this photo is not credible. For one thing, it shows nothing in the background to allow its size or distance to be ascertained. It could as easily be a tiny model seen close-up as a giant hovering craft. In the 1990s the Belgian skeptic Wim van Utrecht showed that the photo could easily be reproduced using a small model. In a 2011 issue of **Tim Printy**'s WebZine **Sunlite**, an article by **Roger Pacquay** notes several inconsistencies about the photo.

Now we have a confession. The Belgian news organization RTL reported that the hoaxer has given his "Mea culpa" and now "lifts the veil": The reporter interviewed "Patrick" in his home (the formerly anonymous hoaxer is now known to be **Patrick Marechal**), where he showed them many slides and prints. "l'OVNI de **Petit-Rechain** n'est pas un vaisseau spatial venu d'une lointaine galaxie mais un panneau de frigolite peint et équipé de trois spots" ("The UFO of Petit-Rechain is not a spaceship from a distant galaxy but a panel of painted styrofoam with three spots affixed.")

"On arrive à tromper tout le monde avec une bête maquette en frigolite". ("One has managed to fool the whole world with a silly model made of styrofoam.")

Leslie Kean ignored this embarrassment for quite a while before finally admitting that she (and many other UFOlogists) had been hoaxed.

# "Jack LeMonde" – A Flying Saucer Photo in 1945!

The 1945 **UFO** photograph known as the "**Jack LeMonde**" photo (a pseudonym meaning "the world") created a lot of interest in "serious" UFOlogical circles when it first surfaced in 2000. It shows a young Marine seated on a horse, taken at a riding stable in **Burbank**, California. Behind him is clearly seen a round metallic-looking object, with a tall and ornate tower protruding. Because "flying saucers" were unknown in 1945, the "blemished" photo was quietly pasted away in a family photo album, its "saucer" only attracting attention much later. UFOlogist John Alexander met with the elderly "Jack LeMonde," and was impressed by his sincerity and modest ways.

However, objections soon began to be raised, suggestg that the "UFO" was just a light fixture suspended on a wire. Alexander held fast, arguing that no supporting wire could be seen in the photo (perhaps not realizing how easy it was for such small details to get washed out), and that the city of Burbank was using different streetlights at the time, anyway. In 2006, a French-language website of the group **RRO** revisited the photograph, showing how an enhancement of contrast does indeed bring out something looking very much like a wire running over to the "UFO". Furthermore, a photo was located showing a hanging streetlamp in nearby Pasadena used since the 1930s, identical to the UFO in the photo.

## 2004: The Mexican Air Force films Infrared UFOs

A now-classic **UFO video** was taken on the afternoon of March 5, 2004 in southern **Mexico**, over the states of Chiapas and **Campeche**. A Merlin C26/A aircraft of SEDNA, the Mexican Secretariat of Defense, was on routine patrol looking for drug smuggling or other illegal activity. It was using the STAR SAFIRE II infrared sensing device manufactured by **FLIR** Systems of Portland, Oregon. From an elevation of 3,500 meters (approximately 11,500 feet), the infrared sensor system recorded a sequence of unidentified objects, at one point numbering as many as eleven.

These "UFOs" (like most "UFOs" photographed) appeared only as bright points of light, showing no detail or structure. But they were different from the run-of-the mill "UFO sightings" because the objects could not be seen visually, appearing only in the infrared images. Infrared systems such as the STAR SAFIRE II detect electro-magnetic radiation in the 3 to 5 micron bandpass, with a resolution of 640 by 480 pixels. Images are formed by the differences in the scene's apparent infrared radiant intensity caused by temperature differences and emissivity differences, and to a lesser extent reflected energy. Thus, objects hotter than their background appear to be self-luminous. The images can be recorded digitally or on conventional video recording equipment (with lower resolution) as was the case here. Infrared systems are useful for daytime operations, especially in humid climates where visibility tends to be poor, because infrared radiation penetrates the atmosphere better than visible light. These objects were recorded as brilliant objects in the infrared, suggesting that they were emitting enormous amounts of heat. However, due to the nature of the video recording and lack of knowledge of the sensitivity parameters, actual temperatures are impossible to ascertain from the available data.

To get a better understanding of the operation of the infrared recording system, and the situation in which it was being employed, I contacted John Lester Miller, author of over forty scientific papers and four textbooks on infrared and electro-optical technology. He is also an active member of Oregonians for Rationality. He explained,

> "The UFOlogists' concerns about not being able to acquire the objects visually is meaningless. These systems are specifically designed to detect objects that cannot be seen by the human eye. Frankly, it would be a waste of taxpayer's money to equip a plane with a system that could not detect objects invisible to the eye. If the eye could see everything that the IR sensor can, then it would be far cheaper and more effective to put a few privates in the aircraft with binoculars. But this isn't the case. By exploiting infrared electromagnetic radiation caused by thermal and emissivity differences in a scene, a different landscape is revealed. For example, infrared imagers can easily detect humans and animals at a distance of several miles at night where the eye or CCD sees nothing but darkness. Moreover, being longer in wavelength, typically Infrared radiation transmits better though the atmosphere than visible and is exactly why it is now being deployed on commercial aircraft for enhanced vision for pilots."

# Bad UFOs

These images were viewed though one of the worst atmospheric conditions possible. Hot, humid, and partly cloudy at a land and sea interface, during the thermal instability of sunset or sunrise. This represents one of the most difficult atmospheric conditions for accurate imaging. These conditions seriously impair the quality of the images in the visible and even reduce the quality in the Infrared. In these stressing atmospherics, it is no surprise that there was nothing visible to the eye and the images are blurred and altered in the infrared. The smaller images below the main images could be reflections from water or ground (common in the infrared) or even mirages. All of these phenomena are typically observed in such conditions. The bending of the light in the atmosphere going though multiple dynamic layers of varying indexes of refractions also call into question the angular indications."

Any representation of a three dimensional scene on a two dimensional surface (be it a painting, photograph, television or Infrared scene on a display) lacks absolute range information. It is impossible to infer the range from the image of an object based on brightness or size, unless the brightness and size are well known, the atmospheric conditions are well known and the sensor settings are known. There are simply too many unknowns to solve the equations. Painters and photographers have long exploited the human predisposition to read range into a two dimensional scene for both optical illusions and stunning artistic effects. Infrared sensors frequently employ a laser rangefinder option for this very reason, which was not present on the sensor that acquired these images. The only way to accurately determine range is by radar.

At some points two unidentified objects were reported to turn up on radar. However the position and number of the radar objects did not even come close to matching that of the infrared ones, so whatever the radar targets were, they were not the same as those recorded on video. This is generally the case when visual sightings of UFOs (or in this instance, IR sightings) are "confirmed" by radar. Unfortunately, no radar data from the aircraft was recorded, so we must rely on the crews' recollection of what it showed. The military radar operator in the city of Carmen was contacted, and it was not showing any unknown objects. UFO researcher Brad Sparks, plotting the direction and distance of the aircraft's radar returns on a map, found that some of them appear to match the position of the Yucatan Highway 186. He suggests that the measured velocity of the radar objects (52 knots, or 60 miles per hour) is quite consistent with the velocity of trucks, and so concludes that some, although not all, of the moving objects spotted on radar are due to trucks on the highway. There are many kinds of objects, both flying and on the ground, that can turn up as targets on aircraft radars and infrared sensors.

The tapes were released to Jaime Maussan, a well-known Mexican broadcaster and UFOlogist who has made a career out of the sensationalist promotion of supposedly 'unexplained mysteries.' To understand the reputation of Maussan in Mexico, imagine a well-known broadcaster with a reputation for sensationalism such as Geraldo Rivera or Jerry Springer, but one who specializes in UFOs. Maussan's pronouncements range from the sensational to the absurd. For example, Maussan spoke at the *Bay Area UFO Expo*, San Jose, California, in September, 2000 about "glowing extraterrestrials" being widely seen in Mexico, and claimed to have sighted one of them himself. He also showed a photo of a supposed alien "life form" reportedly encountered by Apollo 11 astronauts on the moon, labeled "El Hombre de la Luna." If one wanted an objective evaluation of the objects in the video, Maussan would be last person to turn to. Indeed, a May 17, 2004 editorial in the influential Mexican newspaper *La Cronica de Hoy* by Raul Trejo Delarbre suggested exactly that.

On May 11, Maussan held a press conference promoting the videos as a sensational mystery. Maussan's story was widely run by the news media worldwide, including the Associated Press, CNN, Reuters, MSNBC, USA Today, and Fox News. He soon had the videos on his TV show, as well as on his website. The website is filled with a mixture of information and misinformation concerning the objects, in typical Maussan fashion. It claims that the "halos" seen surrounding the objects is evidence of a powerful "magnetic field." It goes on to discourse knowingly about the objects' "frequency" and "vortex", as well as their supposed violation of "entropy", all of which is complete pseudo-scientific balderdash.

Unfortunately, many would-be skeptics leaped to make hasty pronouncements about the objects, thereby giving all UFO skeptics a bad name. The Urania Astronomical Society of the state of Morelos told the newspaper *El Universal* on May 13 that the UFOs filmed might be a group of weather balloons. Dr. Julio Herrera of Mexico's National Autonomous University told the Associated Press that the UFOs were electrical flashes in the atmosphere, a theory that makes very little sense. A few days later, he was attributing them to "ball lightning." Rafael Navarro of that same university told a press conference on May 14 that the UFOs were luminous sparks of plasma energy. Mexican astronomer Jose de la Herrin stated that the stationary objects could be meteor fragments. UFOlogists were soon gleefully mocking these absurd explanations, making it look as if skeptics were ignorant fools who couldn't recognize an alien spacecraft when they saw one. There is nothing wrong with saying "I don't yet have enough information to know what the objects are, but I am confident that when more facts come in, we'll find a prosaic explanation."

By May 20 some skeptics who prefer to remain anonymous had identified the probable source of the objects: burning oil well flares from offshore oil platforms in the Bay of Campeche. This region is the center of Mexico's petroleum industry, containing over 200 wells on nine platforms, many of them close to Ciudad del Carmen, the "city of Carmen." (One of the voices on the video can be heard saying that the objects are "at Carmen".) At that point we thought that there had been some temporary burn-off of excess natural gas within the well – it turns out that the oil well flares burn more or less continuously in this region. They also have large steam generating plants in the area that pump incredible amounts of hot steam deep into the ground to increase the pressure and ease the flow of oil.

One anonymous "concerned outdoorsman" who works on offshore oil platforms wrote on an environmentalist website:

> Each day while I work, I see flares burning at such a rate that it is almost unbelievable to the human eye. I'm told that all gas sources are being burnt off through the flares just to keep the crude oil flowing from each well. Each production platform consists of at least twelve penetrations drilled into the sea floor reaching to different depths. Each platform has a flare some have two, in which are roaring twenty four hours a day, three hundred and sixty five days a year.... At night when looking across the bay of Campeche, it looks like a spotted forest fire out of control in the distant far yonder, in

any direction you choose to look. The black smoke rolls and it never stops!

On May 26, Capt. **Alejandro Franz** of the private Mexican UFO research organization **Alcione**, who is far more skeptical than Maussan and his colleagues, independently came to the same conclusion. A former pilot who has flown extensively in that region, Franz wrote on the widely read *UFO Updates* on-line forum:

> "**Cantarell** Field or Cantarell Complex is the largest oil field in Mexico, located 80 kilometers offshore in the Bay of Campeche... The objects (lights) are in a fixed position with a dark background (the sea) while the camera on board is following the lights that are showing in the screen as a very brilliant source of light... the lights are coming from steady Oil Platform flames (passive fire) located in the Gulf of Mexico between 50 and 90 Km from Ciudad del Carmen City where the objects, at least one light as the FLIR or RADAR operator tells is exactly over Ciudad del Carmen"

On the Alcione website, Franz provides a great deal of information and photos concerning the Cantarel offshore oil wells and their continuous flares. No reasonable person could see his photos comparing the flaming offshore platforms with the "infrared UFOs" from the video and reject the likelihood that the two are the same.

Franz is mistaken in suggesting that the aircraft was headed north at the time that the videos were taken. The aircraft was headed eastward, at an azimuth of approximately 80 degrees. This is confirmed by an event occurring near the end of the half-hour video, 26 minutes in. The crew members are briefly surprised by the image on the STAR SAFIRE II of a large, round object. They zoom in, and realize that it's the moon coming up. "The moon, it's the moon", they laugh. The moon rose at approximately 17:20 from their location, at a geo-azimuth of 75 degrees. Because the azimuth of the STAR SAFIRE II relative to its aircraft mount is reading approximately -5 degrees (just left of straight ahead), that confirms that the aircraft was on a heading of about 80 degrees. It may or may not be significant that the sun was at this time at an azimuth of 260 degrees, directly behind the aircraft.

## Bad UFOs

The STAR SAFIRE II records the altitude and azimuth of the object it is imaging at all times relative to its aircraft mount. The altitude of the "UFOs" is within a degree or two of the horizon, with respect to the aircraft. The crew said that the objects were at the same elevation as us, i.e. on the horizon. The pitch of an aircraft in "level flight" depends on a number of factors, including its airspeed, trim, bank angle, the configuration of its flaps, gear, spoilers, etc. If the aircraft had a 1 to 3 degree pitch upward from its centerline (typical of normal flight), this needs to be figured into the altitude reading from the infrared sensor. It would mean that a "zero elevation" reading indicates that the sensor is pointed below the horizon, when looking backward.

During the main part of the "UFO encounter," the object's azimuth is reading around -134 degrees, approximately the 7:00 position behind the aircraft. When plotted on a map showing the aircraft's position and heading, this points in the direction of the largest oil well platform complex in the Gulf of Campeche. From the video, the objects can be seen to be over water, but one cannot judge the altitude of the objects above the water, or their relative motion with respect to the water. When two brilliant "UFOs" are seen behind fluffy clouds (figure 1), the IR camera is set to the Medium Zoom field of view, giving a field of 3.4 by 2.6 degrees. We are seeing the two main oil well platforms. Soon afterwards, the operator selects the narrowest field of view using the E-zoom feature. The field of view is only .4 by .3 degrees, which makes the objects appear about nine times larger. We see the result ... when individual flames are resolved on each oil platform, revealing nine or more "UFOs". If you compare this frame to the photo found on the Alcione website showing a daytime view of flames on the oil platforms, you will see that they match up quite well.

The objects appear to be moving with respect to the clouds that pass in front of them, giving the objects the illusion of motion. However, the motion of the aircraft with respect to the clouds, as well as the motion of the clouds themselves, causes the highly magnified lights to appear to shift position with respect to the clouds. Since the azimuth of the objects does not change significantly during the time they are being filmed, it is evident that the apparent motion of the objects with respect to the clouds is caused primarily by the motion of the aircraft with respect to clouds, and not by motion of the objects themselves. This situation is analogous to zooming in using a telephoto lens on your video camera, and pointing to a far away mountain. Then, while keeping the camera pointed to the mountain, walk along a treed path. The trees will have apparent angular motion (or optical flow) due to your movement, while the mountain stays approximately still. When overworked, stressed, disoriented and confined to looking at a 3-dimensional scene on a two dimensional display, it is easy for the crew to become confused as to what is moving. All of the "UFOs" recorded by the STAR SAFIRE II are on the left side of the aircraft, towards the Gulf of Mexico, with the great majority of them around −134 degrees.

The UFO proponents participating in the online "UFO Updates" forum, which includes many "leaders" of the UFO movement, laughed off the valid explanation of oil well fires as flippantly as they did the absurd ones. UFO author Ray Stanford scoffed at "Oil rig flares tracked on radar at near the aircraft's altitude", neither of which statement was true: the radar targets were in a different direction entirely, and distant objects near the horizon may well be on the ground. Roswell champion David Rudiak scoffed at "Invisible, flying oil wells", while Alfred Lehmberg suggested that skeptics might as well propose "Soaring lighthouses and gassy pelicans." Others raised the specter of elves, angels, flaming seagulls, etc. Jaime Maussan argued that the flaming oil wells would not have been visible, because they were from 125 to 200 km or more distant. He neglected to **calculate** that from an altitude of 3,500 meters, the **distance to the horizon** is nominally 211 km, and that atmospheric refraction typically extends this distance somewhat, depending on meteorological conditions, as also does the height above the water of the flames themselves.

By their reaction, the 'leaders' of UFOlogy have showed themselves incapable of distinguishing logical thought from illogical, and science from pseudo-science. The lesson of the Mexican Infrared UFO videos illustrates once again the inability of the UFO movement to perform critical thinking.

## 2007 - Attack of the Drones?

Starting in 2007, pictures of weird, spindly-shaped UFOs started to turn up in UFO websites and magazines, usually submitted anonymously.

## Bad UFOs

Looking like a weird cross between a wire basket and a ceiling fan, "UFO drones" started popping up all over the place.

The first such photos have supposedly come from a fellow in Bakersfield, California known only as "Chad." In May of 2007 he submitted a total of six "drone UFO" photos to the *Coast-to-Coast AM* website, where they were posted. He wrote, "my wife and I were on a walk when we noticed a very large, very strange "craft" in the sky... The craft is almost completely silent and moves very quickly... I see this thing VERY often... It is almost totally silent but not quite. It makes kind of "crackling" noises...it moves almost like an insect." The object in the photo had five protruding arms, one being much longer than the others.

Before long, a second set of "Drone UFO" photos was allegedly taken at Lake Tahoe near the Nevada/California line, submitted anonymously to MUFON, and posted on their website. This craft had four arms, with two significantly longer than the others. Soon six more anonymous "drone" photos said to be from Capitola, California were posted to the internet by a person calling himself only "Rajman." This one had some sort of weird "alien writing" on it. A few days later, somebody known only as "Stephen" produced three "drone" photographs supposedly taken at Big Basin Park, not far from Capitola. The object is somewhat distant, and details are hard to see. About ten days later, a guy named "Ty" submitted twelve "drone" photos, supposedly taken at Big Basin Park on the same day as those of "Stephen," and seen by his cycling group. "Ty's" photos are amazingly close-up, allowing one to see every gear and sprocket and spike in clear detail. After that, a few more pictures trickled in from here and there, but the fad for photographing "drone UFOs" seemed to have run its course. Somebody calling himself only "Isaac" wrote a letter explaining how he used to work on a classified project called "Caret," that utilized captured alien technology to produce antigravity. He also produced what purports to be a technical manual, portions of it heavily redacted, showing parts that seem to have come right off a "drone UFO."

In 2008, a woman in London contacted California private investigator T.K. Davis. She wanted to hire him to find out who photographed the "drones," as thus far every photographer has only given a first name. And she doesn't want to be identified, either. She had emailed "Rajman" with some questions. He replied briefly, then closed his email account. So

## UFO Photos and Videos

Davis and his colleague Frankie Dixon were off to Capitola, trying to identify the specific telephone pole seen in the photo. The whole affair was starting to sound like a Humphrey Bogart movie.

Yet another photo of a spiky "drone" from the Netherlands was quickly identified by several readers as a "Waldorf box kite," which indeed does have exactly that spiky shape. Of course, the clear and detailed but anonymous "drone" photos from California are not the result of cracked windshields or kites, but probably were courtesy of Photoshop or similar software. In fact, some computer graphics whizzes have already posted on *YouTube* impressive animated videos of "drone UFOs." (The admittedly-hoaxed videos are also being posted anonymously, leading one to wonder just what is going on!) Seeing is no longer believing, if indeed it ever was. When I first saw a drone UFO photo, I thought that somebody had dismantled an old ceiling fan, and stuck some radio parts on it. But when I saw on *YouTube* an admittedly faked "Drone UFO" looking just like the "real" ones, not just as a still frame, but rotating and flying, I realized that all the Drones are pure CGI. They exist only in Cyberspace. Marc D'Antonio, MUFON's chief photo analyst and one of the few real experts in analysis of questionable UFO videos and photos, explains why these photos are hoaxes on *www.dronehoax.com*.

The well-known (and highly credulous) UFO promoter Linda Moulton Howe told the MUFON Symposium in 2011 about UFO "drones." Somebody Howe was working with managed to supposedly decipher some symbols along a drone's wing, starting with a number 7. They then Googled that string, which brought up a network address on some internal NASA page from the Clementine lunar orbiter project. This is perfect for a nice conspiracy, because Clementine was a joint project between the Strategic Defense Initiative Organization ("Star Wars") and NASA. Better still, Clementine failed after completing only part of its mission - but maybe it didn't fail, and it was actually spying on alien activity? So Howe was suggesting that the Drones had something to do with Clementine, although she ultimately determined that the "7" wasn't really a "7", or any of the others supposed ASCII-represented characters. They were alien symbols. So they googled the character string that wasn't a character string, and found the correct Conspiracy, anyway.

Then somebody sent Howe (again anonymously) some peculiar-looking documents purportedly from *Project Caret,* a supposed alien research project being carried out in top secret in Palo Alto, CA. It also is drone-related. It purports to tell of several different alien "sources" (races) behind all the UFO activity. I am really impressed with the effort that the hoaxer did to create all this stuff. Since all of the Drone-related photos and documents originate from northern and central California, I suspect that just one individual is behind the entire Drone/Caret hoax. Just one very talented, very patient, and very bored individual. Howe closed her talk with an extremely implausible claim from a man who says he checked out the 1962 book *Ancient Greek Gods and Lore Revisited* from a library, and who was visited soon afterward by Government agents demanding that he surrender the book. Apparently the ancient Greek gods had something to do with aliens, and this all needs to be covered up. It's even more dubious because the only pages I can find mentioning that book also mention Howe and Drone UFOs, so it looks like government agents have been so successful in covering it up that this book on Greek myths no longer exists (if indeed it ever did, which I doubt).

On Sept. 10, 2009 *The Telegraph* of London published a strange photo with a story titled, "UFO or pterodactyl over Argentinian lake? A strange object photographed over a lake in Argentina has been described as either a flying saucer or a flying dinosaur." The somewhat blurry photo, taken using a cell phone, shows a round object with five arms or spikes protruding from it, causing anyone who has been watching the above carnival to immediately exclaim, "It's a drone!" The photo was taken by one **Rafael Pino** (at least this man has a first and last name!), who says he was driving his truck when he spotted the object, and stopped to snap three photos. However, one alert reader down in Argentina wrote, "It does look like a windshield cracked by a rock." An analysis of these photos on the blog *forgetomori.com,* whose motto is "Extraordinary Claims, Ordinary Investigations," suggests that "Indeed, the 'UFO' is apparently in the same perspective in all photos, as if it didn't really move. Note that in the second photo, the line of horizon is tilted... but the UFO's rightmost 'spike', which is actually a crack, is still parallel to it. So, a cracked windshield looks like a good and obvious explanation. "

## 2011: "Miracle UFO" above Jerusalem's Dome of the Rock?

## UFO Photos and Videos

It took about two or three days for the sensational UFO videos from Jerusalem, Israel make their way to the Big Time (the mainstream media). The two videos (originally) appear to show a bright nocturnal UFO flying and hovering over the Dome of the Rock on Temple Mount, the point where, Moslems believe, the Prophet Mohammed took his "night journey" into the Heavens on the back of a horse with the face of a woman (al-Miraj), passing through the Seven Spheres to meet with Allah himself.

For some time, I have been pointing out the lack of consistent UFO videos or photos taken by multiple independent witnesses, who had not been in contact before the incident. That is what we appear to have here, unless the case is a hoax (widely suspected, but not demonstrated).

The first video was posted to YouTube on January 28, 2011 by a user called "eligael." It has a length of 1:45. It purports to show a UFO hovering over the Dome of the Rock in the far distance. It descends quickly to hover just above the dome, then disappears with a flash some 23 seconds later. Two men are heard discussing it, presumably in Hebrew. One man is seen in the foreground (barely), and appears to be filming it with a cell phone camera. According to a video posted to YouTube on Jan. 30 by **HOAXKiller1** titled "HOAX - UFO Over Temple Mount in Jerusalem - Motion Tracked," claims that there is a "parallax problem" where two lines that ought to remain parallel despite small camera motions, don't. Unfortunately, the video is so dark that this is extremely difficult to see. A second video from this same perspective was posted by Eligael on Jan. 29. It has a length of 1:30. It appears to be consistent with the first.

A third video posted to YouTube on Jan. 30 by **50nFit** seems to show the same UFO from a much closer perspective. It has a  length of 0:48, and features people commenting in English. One woman with a thick Southern accent says, "We've seen 'em in Mississippi like this, but never like that." According to "Mr. Mask" on the ***Above Top Secret*** thread, "The third video (though mistakenly called "second" in its title) was of the UFO from a totally different perspective and with what appeared to be many voices. One was a lady from Mississippi. The video was proven to be a moving computer effect added over a still picture easily found [on Wikipedia commons] . The clip is clearly a Hoax and a bad one at that. Truly the work of a desperate and loathsome huckster." "Eligael" posted yet a third video to YouTube on February 1, titled "Another night of ufo over Jerusalem - old city," This one has a length of 1:25. It appears to show a single point of light against a dark sky, dropping down, then flying off mostly horizontally to the right. Nothing is seen but the UFO.

A "fourth" video, also in Hebrew, was posted to YouTube on Feb. 1 by Disclosur3. It has a length of 1:15. It shows a UFO hovering over the dome, looking brighter than in the other videos. The UFO and Conspiracy Website **Above Top Secret** warns, "Due to the possible connection of these clips with a known Hoaxer who is banned from this site, this thread may be 404ed/erased at anytime or thrown in the hoax forum if it is proven that this hoaxer is behind this event."

Whoever is responsible for this hoax, it clearly shows that at least a few people are willing to put far more effort into making a UFO hoax than ever before. "Classic" UFO photo hoaxers such as Mr. & Mrs. Trent, the Lucci brothers, and Rex Heflin, probably invested no more than about an hour or two making their now-famous creations. If "Eligael" is indeed the mastermind behind these clever videos, it is obvious that he has been willing to spend hours or even days creating his masterpiece.

"Why bother to investigate this?", the philosopher might ask. "The burden of proof is on the person making the claim. Unless it can be shown that there is practically no way such a video can be faked, it doesn't prove anything, and so it doesn't need to be debunked." True enough, but when dealing with the public, it helps to be able to go from "this video proves nothing" to "this video is shown to be a hoax."

We can now state with great confidence that the original video, seen above and posted to YouTube on January 28, 2011 by user "eligael," is a hoax. Those who are familiar with video editing software, point out the effects of the digital processing software, proof that the video clip has been through a program to modify it from its original state. Using such a program, the knowledgeable user can insert more or less anything into the video, and make it look at least somewhat convincing.

The proof is fairly easily seen the the video posted to YouTube by user **HOAXKiller1** on Feb. 5. This user had posted several earlier videos attempting to make this point, but they were not so useful in showing exactly what the problem was. But with this video, I think that HOAXKiller has succeeded, showing quite clearly the effects from the editing.

The hoaxer has used something called "Motion Tile" effects in the processing of this video. An artificial camera shake is introduced, to make it look like the video was taken using a hand-held camera. But what do you show, say at the extreme left edge, when the camera moves to the left? You don't have anything at all to the left of the left edge of the picture (although you would if the camera were really shaking). I'm reminded of a time when I was singing in a choral rehearsal of Puccini's *Tosca* in a small opera house. In the Second Act, the chorus is offstage singing a religious hymn, when the villain Scarpia suddenly slams the window, cutting off the sound. One bass had a problem: "What do we do if he's late closing the window?" (A reasonable concern in live theater, perhaps). The director smiled and said, "Then *you* keep singing." The joke is, of course, that in the score the choral part ends at that point, so even if Scarpia doesn't close the window, you still have nothing to sing.

You might solve the problem of "nothing to show" by filling in with blank space, but that would not look realistic. So a "mirroring" effect is generally used. You create data beyond the edge of the frame by mirroring data at the edge of the frame, along all four sides. Watch HOAXKiller1's video above, and you will see this happening. First, we are clearly shown where the mirroring occurs, in the first minute of the video. Next, we see an excerpt from an instructional video for using a video-editing program. It is a tutorial involved with adding or removing camera shake. It explains how to use "motion tiles," with "mirrored edges." This same process was used during the creation of the original Jerusalem UFO video.

This proves that the video did not go directly from the camera to YouTube, but made a stop in between inside a sophisticated video editing software suite. Which is obviously where it picked up its image of the "UFO."

## 2011: UFO Mothership & Fleet over London

Another implausible video showing UFOs has gone viral, spawning a flurry of news stories in the major media. For example, the *Daily Mail* in London asks (June 29, 2011), "Are aliens getting less camera shy? UFOs filmed above BBC building in London." The *Huffington Post* also reported on June 29, "London UFOs: Multiple People Capture Odd Occurrence Over British City" (although not at the same place and time. ). The video that started it all, was posted by a photographer known only as **alymc01**. We see a lot of seemingly random (and pointless) filming of people on the street. The camera points skyward, and we see some small UFOs, and finally a big one (the "Mothership") moving in and out of the clouds.

One has to admit that this looks pretty cheesy. The big "Mothership" looks a bit like a lens flare, but it does not act like a lens flare, its movement unrelated to that of the camera. At first I thought that the small UFOs were birds, but on closer examination they appear to be generated artifacts as well. Actually, that is Alymc01's second UFO video. His earlier video doesn't look nearly as impressive, so it was largely ignored.

There was a great deal of commentary that did not go beyond comments like 'this looks like a computer-generated fake'. And strictly speaking, that's enough. After all, *the burden of proof is not on the skeptic to show that a video is fake. The burden of proof is on someone who claims it shows unknown crafts, to rule out all prosaic explanations.* And since the film was posted anonymously, this makes it all the more dubious. Using the terminology of Mythbusters, that is enough to call this video "busted." But also in the spirit of Mythbusters, let's not stop there. Let's see if we can really blow this thing apart.

British UFOlogist Nick Pope didn't buy it. That's bad for this video, since Pope, who was then on tour to promote the DVD release of the Hollywood space alien movie *Battle: Los Angeles,* buys a lot of dicey things. But apparently this video looks unimpressive even to him. Interestingly, Pope adds, "The slightly suspicious thing, though, is it's a part of London where it just so happens that a large number of film companies and visual effects companies are based. And some of the people do look a little bit self-satisfied. So I suspect this is a CGI hoax, and that someone is showcasing their skills." A perceptive comment!

Surprisingly, *the most useful commentary on this video was found on the UFO and conspiracy-oriented website,* **Above Top Secret.** The forum participants, mostly anonymous, dug deeply and turned up facts that the 'experts' seem to have overlooked. Of course, many of the comments in that very long thread are credulous and foolish, and I don't want to imply that all of the participants are credible researchers. But I am definitely impressed with a few of them!

"**C-Buzz**" commented "100% CGI. 1:18 - 1:22 the object doesn't actually go behind the clouds, it fades out. Not only that it looks like he stuffed up creating this animation because if you have a look at the bottom left there is actually a lighting effect which probably isn't supposed to be there & a RED orb moving across the building." It's hard to see, but it's there. There's also a brief "green flash" on the building, as well as a suspicious-looking red color on the "mothership." I'm not enough of an expert on digital processing to know what this means, but it reeks of digital tampering. Sharp eyes, C-Buzz!

"**LiveEquation**" posts "The video is a scam right and I have evidence. If you start watching the video at 1:21 you will see two artificial bubble glares and then delay of the UFO glare. The UFO vanishes into the clouds first. Then you see 2 fake bubble glares and the ufo glare moving in the same direction *after* the UFO has already vanished, delay of about 1 second. It's actually weird that the UFO cast a glare. That's a giveway. The person who made the video doesn't know jack about optics."

"**GiftOfProphecy**" adds "This video is clearly fake. You can prove it by watching the video starting at 1:00 and after, and stabilizing the video. You can see they did a horrible job motion tracking the camera movement... probably because they have a rolling shutter camera. If you watch the UFO you can see it is not shaking with the camera perfectly, it is shaking independently. However the "UFO" is shaking the same rate and nearly the same magnitude, it's direction and position are just not synchronized. That to me indicates several bad motion track points. In order to insert a fake UFO into the video they had to track certain pixels as they move and shake around, then apply that tracking to the UFO so it moves exactly the same as the camera (match moving). Sometimes the pixels will move say 10 pixels in one direction, yet the computer detected the pixels move 12 pixels, and that creates a bad tracking point. Normally you can fix bad tracking points by hand, but when there is about 30 tracking points per second, it becomes very time consuming. If you apply the motion tracks to the UFO when it has bad track points, it will wobble and shake around similar to what you see in the video."

"**charlyv**" noted "Fake, stop action in frames shows no motion blur, Impossible for such recorded speeds in any consumer digital camera, regardless of make or resolution."

Then "**davespanners**" opened up a whole new angle of investigation: "This is filmed outside coral bookmakers in clipstone / great portland street in London. If you google search that building You will eventually find this page, which is a tv production company that is in the very same building. From their web site:

> The Mill creates pioneering visual effects for the advertising, music, television and film industries. We craft commercials, music videos and generate compelling film and TV. We build installations, projections, applications and create multi-media content and experiences.

"**EnigmaAgent**" replies with a photo of Managing Director Mike Smallwood, taken from that company's website, who appears to be the same guy seen smiling in the video, apparently enjoying this incident 'way too much.

"Heliocentric" dug further, and found a link from the The Mill's website to a particular commercial for Sony. And that same Sony commercial is a "favorite" on the YouTube page of Alymc01, who photographed the "UFOs." The noose tightens!

Chillingly, "GiftOfProphecy" observes that video hoaxers are now using claims of "copyright infringement" to make YouTube remove videos showing how an original video was hoaxed: "the hoaxer '50nFit' is claiming copyright infringement on the video that proves his Jerusalem video is a hoax... Now the videos that prove his London UFOs are a hoax were removed to avoid complete suspension [of his YouTube account]. The "HOAXKiller1" channel may be suspended anyway because YouTube doesn't understand Fair Use laws, and allows the deceptive scumbag hoaxers to retaliate and claim copyright on videos that are for research and analysis." (As of this writing, "HoaxKiller 1" is still operating on

YouTube, as is HoaxKillerFriend) In other words, if you place a video on YouTube showing how a UFO video was faked, the hoaxer might contact YouTube to force you to remove your analysis, claiming "copyright infringement."

## 2012: An Olympic UFO

In some videos of the opening ceremonies of the 2012 London Olympics, an unusual object is seen. It appears to be glowing, and perhaps transparent. An article by Tom Rose in T*he Examiner (since removed)* attracted a lot of attention. The article references a YouTube video:

> The above video is "second-hand," taken from a broadcast. So I would not trust it very much. However, the following video posted to YouTube does appear to be first-hand, that is, directly from the news video. And it does indeed show the object, so I will assume that the object has not been added to the video or hoaxed in any way. So what is it?

It seemed likely that the object is the Goodyear blimp. As commenter Chris Jisi posted to *the Examiner* article above,

> It's the Goodyear Blimp Spirit of Europe II, an American-made 130-foot-long "Lighship," which is lit by an internal bulb. With the Goodyear banner removed, as per the rules of the Olympics about no commercial advertising, it could look very much like a UFO.

This sounds very convincing, but can we find any additional support for it? Just take a look at the photos from the British website of the **Goodyear Blimp** on my **BadUFOs** Blog, some of which are uploaded by Mobile users. I think those people know their own Blimp when they see it. The resemblance between that object, and the unknown object in the video is obvious.

**Case Closed: The object was the Goodyear Blimp.**

## 2012: Flying Saucer, or Fly: The case UFO skeptics have been dreading?

## Bad UFOs

On March 13, 2012 the well-known UFOlogist Leslie Kean wrote a story for the *Huffington Post* titled "UFO Caught On Tape Over Santiago Air Base" **Kean** is the author of the 2011 New York Times bestseller *UFOs: Generals, Pilots and Government Officials Go On the Record* (see my review "'Unexplained' Cases—Only If You Ignore All Explanations" , *Skeptical Inquirer, March/April,* 2011, now on-line at **debunker.com**) The article described a supposed UFO caught on video during a November 5, 2010 air show in Chile (although nobody saw the supposed UFO until it was spotted in the video afterward). The video was provided by CEFAA, (Comité de Estudios de Fenómenos Aéreos Anómalos) the Chilean Air Force's equivalent  of the now-terminated U.S. Air Force's Project Bluebook. The video was said to have been subjected to "intense scrutiny" by a panel of "scientists from many disciplines, aeronautical experts, and air force and army photogrametric technicians." They found it unexplainable. (Kean is very impressed by panels of supposed "experts," no matter what foolish things they might say, like the French COMETA, which fellow UFO proponent John Alexander called "an embarrassment... unsubstantiated data from questionable sources.") ***She asks provocatively, "Is this the case UFO skeptics have been dreading?"*** She obviously thought she had an extremely strong UFO case here, and provided a video, exactly as released by the Chilean CEFAA (since made "private").

Kean claimed that the object was captured on seven different videos, by seven different people, each from a different vantage point. However, the video released seems to represent just one of the seven videos, and even that one video is not complete, but instead we see only some carefully chosen snippets. (She later cautioned, "You can't draw conclusions from looking at only one tape."). Plus, the video was released only in a highly compressed 10 frames-per-second version, instead of the standard 24 fps. Somebody seemed to be "stage managing" this case to let us see only certain portions of the "evidence."

Kean said, "Each video included three different, mainly horizontal loops flown by the UFO within seconds of each other. The object made elliptical passes either near or around each of three sets of performing jets. It flew past the Halcones, F5s and F16s at speeds so fast it was not noticed by the pilots or anyone on the ground below." She continued, "This extraordinary machine was flying at velocities too high to be man-made.

## UFO Photos and Videos

Scientists have estimated the speed, depending on the size of the object, to be at least 4000 - 6000 mph." She should have realized that speed estimates are only valid if one knows for certain the distance from the "UFO" to the camera, which in this case we obviously don't.

Kean continued, "Images show it as a dome-shaped, flat-bottomed object with no visible means of propulsion. The rounded top reflects the sun and appears metallic; the bottom is darker and flat, emitting some form of energy which is visible in photo analysis. Infrared studies show the entire object is radiating heat, just like the jets." She did not explain how it would be possible to do an "infrared study" of the object captured by an ordinary video camera with no infrared capabilities. She seemed to not even realize that this claim is absurd.

Almost immediately, I am happy to say, many people began critically examining this supposedly game-changing video and its "expert analysis," even in places like the conspiracy-oriented *Above Top Secret* forum. Many noted the object's resemblance to that of an insect flying wildly close to the camera. Poster **UFOglobe** compared the close-up of the "UFO" in **Kean**'s article with that of bees in a swarm . The resemblance is easily seen.

A detailed discussion and analysis of the CEFAA video appears on the website of The Hoax Killer, who did excellent work exposing the recent Jerusalem UFO video hoaxes. He downloaded the highest-resolution version of the video from the Huffington Post site. He writes, "After studying the videos very closely I discovered the objects are passing in front of the hills a couple times. I was also looking for other insects that might be flying around closer to the ground and found what I think is one that is visible for moment." He posted a short 30-second video to YouTube, highlighting the "UFO" as it is seen not only flying in front of the distant hills, but *in front of the nearby ground. **Case Closed: this "UFO" is a flying insect.***

Skeptic Tim Printy notes, "What this demonstrates is that the original video that was posted was edited in a way so one could not see the UFOs with the ground in the background." In other words, somebody - presumably in the CEFAA - edited the video with the deliberate intent to deceive. Otherwise, it would have been obvious that this is a bug. Blogger

## Bad UFOs

**Kentaro Mori** noted how the CEFAA recently released another supposed UFO photo touted as the "**best evidence**" yet for a UFO, but was almost certainly a light reflection. He noted that the **CEFAA** was then, as now, unwilling to release any additional images or information. Clearly, any future UFO claims from Chile's CEFAA must be viewed with great suspicion.

As one might expect, this case was widely discussed on the Internet in the weeks afterward. Even many who are favorably inclined toward UFO claims dismissed the El Bosque video as a flying insect. But instead of giving up and admitting the obvious, on March 24 Leslie Kean doubled down on the case, writing on her Facebook page, "Have patience folks. More info will be presented soon. You can't draw conclusions from looking at only one tape. And please remember, these images have been analyzed by experts in Chile. Didn't I emphasize that enough in my Huffington Post story? They too thought the footage was a bug at first, until they collected the other tapes. It is an insult for you carry on about this being a bug, when obviously if you can figure that out so easily, the Chilean experts would have done the same."

Like I said, Kean *really* is trusting of so-called "experts" who make proclamations about UFOs. She seems to have forgotten that various "experts," many with PhDs, also analyzed the Petit-Rechain UFO photo from Belgium, and concluded, "the picture was not faked" (Kean 2010 p. 30*).* Yet it's now a confessed hoax. It seems that "experts" do not have a very good track record when it comes to "authenticating" a UFO photo or video.

But the story doesn't end here. YouTube user **Stiverinmypocket** analyzed a longer video of the same November 5, 2010 airshow on the Chilean website *Aviacion Total*, independent of the video of the CEFAA. It, too, is filled with fly-like UFOs buzzing in front of the airplanes. However, on this video the "UFO" moves in an opposite direction with respect to the jets than in the CEFAA video. Stiverinmypocket wrote, "My intension was to show that the object in the CEFAA video can NOT be a UFO, because the respective object in this old video travels in the opposite direction. It is probably another bug close to Olave's camera, who was shooting simultaneously from another vantage point." I'm sure that there are plenty of bugs to go around.

## UFO Photos and Videos

In a March 25, 2012 posting on his **Ufo-Blog.com**, Michael Naisbitt noted that there is a March, 2010 video of an air show at El Bosque, eight months before the now-famous one, showing similar fly-like UFOs zipping around. So either UFOs infest the **El Bosque** airfield in Chile, or else flies do. Perhaps its name should be changed to El Mosca airfield – "The Fly."

"Is this the case UFO skeptics have been dreading?" asked Leslie Kean, hopefully. As a skeptic, I don't "dread" any such thing. I would welcome *solid* evidence that something extraordinary is flying around openly in Earth's skies. But it's been said many times that extraordinary claims require extraordinary proof, and UFO proponents will have to present a far better video than a fly buzzing around to prove their case. Indeed, the very fact that a video of a fly doing loops is being cited by some of the world's top UFOlogists as among the best UFO images of all time reveals how utterly lightweight even the "best" UFO photos and videos are.

On the afternoon of April 13, Leslie Kean finally posted to the Huffington Post her promised update on the highly-controversial video from El Bosque Arifield in Chile, exactly one month after her initial story about it. Strangely, unlike Kean's initial story ("Is this the case UFO skeptics have been dreading?"), there does not seem to be any link to the update on the Huffington Post home page. However, the update appears prominently on Kean's Facebook page. It almost seems that she does not want to bring any new readers into this controversy, and is writing only to maintain credibility with those already involved. (I suspect at this point Kean wishes she had never heard of the Chilean Air Force UFO group CEFAA, but having embraced this Tar Baby, she is unwilling to admit that her new dress is covered with tar.)

Her new piece is titled "Update on Chilean UFO Videos: Getting the Bugs Out." Surprisingly, this update changes almost nothing: we don't really learn anything that we didn't know before. She quotes Alberto Vergara, "an expert in digital imaging," who stated that "When we examine the whole scene frame by frame, we have been able to realize that [the object] has, apparently, moved at a speed far superior to any flying object of known manufacture." Neither Kean nor Vergara explain how he could possibly know the speed of the object without knowing how far it is from the camera. But Vergara is an "expert," so Kean doesn't question this obvious absurdity.

## Bad UFOs

Kean complains that "Skeptics caused quite a stir by taking it upon themselves to do their own "analysis" of the video clips and then to declare, with bravado, that the object of concern was simply a bug. Often this involved misquoting or misrepresenting me and the CEFAA in accompanying text." [Kean does not specify what supposed "misquotes" or "misrepresentations" she is referring to]. "The question of qualifications aside [we skeptics, you see, are not "qualified" to analyze these videos, but somebody like Vergara is], these individuals were handicapped by one even more overwhelming problem: They were working without the necessary data required to make a proper analysis, and, most importantly, they were looking at video clips pulled from only one of the multiple cameras."

This is a very strange complaint: if people are "working without the necessary data," it is because the CEFAA refused to release any more data (although in reality, the clips from the single video already released contain plenty of information to conclude the "UFO" is an insect). So she blames investigators for looking into this case prematurely (a case she suggested was "the case UFO skeptics have been dreading"), rather than blaming the CEFAA for being secretive. And people "were looking at video clips pulled from only one of the multiple cameras" for a very good reason: the CEFAA has only released video clips from one camera, and people cannot analyze what they're not allowed to see.

"In accordance with the wishes of the scientific team in Chile and these new analysts, General Bermúdez will not be releasing any more videos now, so that the public can be fully informed and maximum understanding achieved when the full package is released. Those involved agree that the new studies should be completed first." In other words, the message to those who want to investigate this high-profile case is: sit still, shut up, and we'll let you know when our "experts" have all of the answers for you.

Then Leslie Kean gets into a discussion of beetles, largely, I suspect, to deflect attention from flies. She presents some pretty good arguments to suggest that the object in the video probably isn't a beetle. Beetles fly more clumsily than the object we see. That's why I think that the insect in the video is probably a fly.

## UFO Photos and Videos

Interestingly, the UFOlogist A. Gevaert in Brazil reports "the two major and oldest official UFO research organizations in South American, one from Uruguay (founded in 1979) and other from Chile ([CEFAA] founded in 1997), have decided to establish a cooperation agreement to work together to both investigate new cases, to evaluate new and old cases and to promote Ufology in general among the scientific community of all South America, but, of course, concentrated in both countries." So it appears that, in Chile and Uruguay at least, the government-sponsored UFO investigative organizations are trying to strongly promote UFO belief. Additionally, Alejandro Rojas of *Open Minds* magazine reported on December 16, 2014 that France would be joining them:

> The Aeronautical and Astronautical Association of France (3AF) has signed an agreement with Chile's Committee for the Studies of Anomalous Aerial Phenomena (CEFAA) to cooperate on the study of UFOs. 3AF posted an article on their website late last month explaining their newly founded relationship with CEFAA.

> 3AF is an important organization in France, and in Europe's aeronautical industry in general. Similar to the United State's American Institute of Aeronautics and Astronautics [AIAA], it is a society for those participating in, or interested in, the aerospace industry.

## UFO Hoaxes? There's an App for That!

Anyone who follows news stories about UFOs knows that many if not most of them involve photographs. Frank Warren of *The UFO Chronicles* has written an eye-opening article (June 22, 2013), "**UFO Hoaxes** with the Touch of a Finger" . Warren is well known as a UFO proponent, but he is no friend to hoaxers. I knew that it was possible to create all manner of digital UFOs in photographs. What I did not realize was how easy it had become. Warren notes that there are many Iphone and/or Android **apps** written *specifically for inserting UFOs into photos*:

UFO Camera gold

UFO Photo Prank (also sometimes called UFO Revelator, or OVNICA).

## Bad UFOs

UFO Camera

UFO Photo Bomb

Camera 360

*The author took a photo of the Unarius society's headquarters in El Cajon, California, then used an Android App to add UFOs and an alien.*

Most of these Apps allow you to select the UFO you wish to add, size it, and place it where on the photo you want it to be. One of the UFO choices in UFO Camera Gold is my very own Cottage Cheese Container UFO, which can be seen on the cover of this book, as well as on my Skeptics' UFO Page (*http://debunker.com/ufo.html*). Some of these Apps can insert aliens into your photos, as well. Other Apps can add an assortment of ghosts in your photos, wherever you want them, with a selectable degree of transparency.

Warren cites two examples of credulous UFO news stories that have been written about fake photos made with Apps just like these. It's true that these are not great-quality fakes. Warren says, "For most seasoned Ufologists the hoaxed photos are blatantly obvious; unfortunately, that minority won't stop the MSM [Main Stream Media] from paying heed to the latest hokum produced." Unfortunately, he is quite correct: it seems that some of the most credulous people around are reporters, who are supposed to be skeptical by their profession. I suspect that the cynical pursuit of sensationalism and ratings is really what is behind that.

What all this means is that it is now trivially easy for just about anyone to produce a semi-convincing UFO photo hoax. And since "progress" in software is inevitable, we can expect to see better and better UFO hoax photos with each passing year. Which means: unless you can absolutely confirm a photo's origin, and confirm that it was not simply added using hoaxing software, you can't believe anything that you see in a supposed UFO photo any longer.

## Alien Peeping Toms and Stan Romanek

On his website http://www.stanromanek.com/ he claims "The Stan Romanek case is the most scientifically documented human/extraterrestrial encounters in the world. There have been well over 100 individually unique experiences that Stan has encountered since December, 2000 that remain unexplainable." A man of many talents, Romanek notes that "Being abducted is only one of many experiences," lest you think that he just sits and waits passively for his extraterrestrial pals. "There have been hundreds of witnesses for dozens of events that defy our current understanding of reality. Witnesses, photographs, videotapes, physical evidence, police reports and scientific analysis have confirmed the validity of these experiences beyond reasonable doubt." Exactly where this "evidence" can be found is not stated; presumably we'll just have to wait for the documentary. "Scientists from top universities have been analyzing various aspects of this case for several years with amazing results, expected to be presented to the public and scientific community in the near future." Since none of these alleged "scientists" are named, or their universities, you can draw your own conclusions.

Romanek's story concerning the so-called "boo video" of the peeping alien is this: in 2003 he was living in Nebraska, and was allegedly troubled by a Peeping Tom who was believed to be spying on his teenage daughters. So he set up a video camera pointing out the window, said to be 8 feet off the ground, in hopes of capturing an image of whoever might be doing this. When he played back the video, he found what he had captured was not some neighborhood pervert, but instead E.T. Somehow Romanek managed to keep the entire amazing story out of the news for five years, until 2008.

Actually, Romanek was not the first to claim that extraterrestrials were peeping into his windows. As I reported in my *Psychic Vibrations* column in the Fall of 1978 , the famous UFO abductee Betty Hill claimed that "window-peeping flying saucers sometimes fly from house to house late at night in New England, shine lights in the windows, and then move on when the occupants wake up and turn on the lights." A year later, we reported about the UFO research of John Brent Musgrave, of Edmonton, Alberta, who received a $6,000 grant from an agency of the Canadian government to support his research into UFO sightings in Canada. This enabled him to compile a "Catalogue of Occupants and Critters," which he found fell into eight types. Among the types are "peeping toms and molesters." Nonetheless, window-peeping aliens had not been reported *recently*, at least not until Mr. Romanek's video.

On May 29, 2008, skeptic Bryan Bonner of the Rocky Mountain Paranormal Research Society got together with five members of the society to try to replicate Romanek's "Boo Video." They started working about 8 PM Thursday night. "We rented ourselves a 4-foot-tall foam latex alien," he said. "We were going to buy one, but I didn't want to blow the $230." They used video 3-D animation graphics to make its eyes move. "What they're claiming would take thousands of dollars and a lot of time ... we pulled the whole thing off for $90 and in five or six hours."

Within a few days, *Youtube* was offering several "enhanced" versions supposedly representing the Romanek video, some with the "alien" singing or dancing. Hardly anybody, apart from true-believing dedicated UFO buffs, seems to have looked at Romanek's "Boo Video" without laughing.

But Romanek's "evidence" just kept getting better. Later Romanek's ETs were replaced by as many as nine alien hybrid little girls, who intrude upon his telephone calls, and also play now-you-see-me-now-you-don't. One of them is said to be his wife Lisa's daughter from a previous UFO abduction. Stan gets a few not-quite-clear photos of strange-looking little girls, whose images probably have been Photoshopped to give them ET features.

Stan says he injured himself falling off a ladder, and was going to get corrective surgery. However, before the operation he was abducted by ETs, and his injury was miraculously healed, to the astonishment of his doctor. Speaking at the *International UFO Congress* in 2012, Romanek mentioned in passing that someone had anonymously mailed him actual photos of the true Roswell crash debris, and he flashed them tantalizingly on the screen. Stan Romanek is a one-man paranormal factory. After his talk, Romanek was at his table in the vendors' room (he and his wife by then had three books of wild UFO claims), I introduced myself to him, and gave him my "Bad UFOs" card. I asked him why the aliens were following him around. We chatted very briefly when somebody (probably his wife) apparently whispered, "don't talk to that guy." Suddenly it was, "I can't talk to you. You just bad-mouth people. Go away." I attempted to get a photo of him and his wife (others were doing so), but he turned his face away from me (how I now regret not getting that image of Romanek avoiding the camera!). He said he'd have his lawyer sue me if I took a photo; I replied that was ridiculous, since he was a public figure in a public forum. So much for "UFO research!"

**Stan Romanek**'s credibility suffered a serious blow when on Feb. 13, 2014, he was **arrested** on charges of possessing and distributing **child pornography**. He was released that same day on a $20,000 personal recognizance bond. He protested his innocence, telling the police that "the U.S. Government is conspiring" against him. Soon afterward, on February 24, Romanek claimed to have been assaulted when he went out to his mailbox to pick up the mail, and photos of the injuries he says he sustained were widely circulated on the internet. However, a police investigation concluded that the evidence was "not consistent with a fight having occurred," and the investigation of the alleged assault was suspended. As of this writing, he is awaiting trial.

# 4 UFO CRASHES AND RETRIEVALS

*Frank Scully's book about a supposed 'saucer crash' in Aztec, NM.*

A large part of contemporary "Science Fiction" UFOlogy consists of promoting one or more alleged UFO crashes. The first alleged UFO crash to gain widespread attention was in the 1950 book *Behind the Flying Saucers* by *Variety* columnist Frank Scully (1892-1964). Based upon the tales told by Silas Newton (1887-1972) and Leo GeBauer (said to be an inventor and a government scientist, respectively), it told of a saucer allegedly crashing near Aztec, New Mexico, in 1948. Among its many dubious claims was that a saucer 99.99 feet in diameter (the saucers allegedly followed "the rule of 9s", when measured in multiples of a long-dead British monarch's foot) crashed in Hart Canyon, containing the bodies of sixteen little men, dressed in the clothing styles of 1890. (*Behind the Flying Saucers* is available as an E-book from *scribd.com*).

But in 1952 San Francisco newspaperman J.P. Cahn convinced the editor of *True* magazine that the Scully saucer crash story was either the greatest event of modern times, or the greatest hoax. The editor agreed, and the result was a long investigation culminating in two lengthy and devastating articles in *True* Magazine. Cahn proved that Scully's sources, Silas Newton and Leo GeBauer, were con men who made a living swindling people by claiming to have devices for finding oil and minerals in the ground. Newton's alleged sample of strange extraterrestrial metal was in fact plain aluminum.

In fact, so convincing was Cahn's demolition of the Newton/Gebauer story, that claims of "saucer crashes" were ignored until at least the mid 1970s. For decades such claims all but disappeared. But after twenty years or so, the stench of the hoax in Scully's book had faded somewhat, and crashed saucer tales began to reappear. The "Aztec crash" story was picked up again in the 1970s by Robert S. Carr (1909-1994), who claimed to be a professor and always called himself "Dr Carr", even though he had never attended college. Carr's son Tim has written an article exposing the tall tales of his late father (*Skeptical Inquirer*, Summer 1997), but no one in Aztec seems to care. Veteran UFOlogist Leonard Stringfield (1920-1994) began collecting crashed saucer stories, and by the late 1970s was writing papers about *Retrievals of the Third Kind*. However, Stringfield never had any proof for his claims, and his crashed saucer stories were little-known to the general public. In 1987, UFO researcher William Steinman, with co-author Wendelle Stevens, published *UFO Crash at Aztec*, trying to re-legitimize that claim. It got very little "respect" - even Stanton Friedman didn't buy it.

Since aliens have been such good business for Roswell, which recently recorded 192,000 visitors to its UFO Museum, the good folks in **Aztec, New Mexico** were probably wondering how they might tap in to the extraterrestrial cash cow. The town planned a "50th Anniversary Celebration" on May 25, 1998 to commemorate the date of its own supposed **UFO crash** in 1948 (although the crash date is also variously given as February 13, or March 25, not that it matters). A first annual UFO Conference was held to benefit the Aztec Public Library. The last such conference was held in 2011; there wasn't enough interest to continue. The Aztec crash seemed to be a hoax, and that was the end of the story.

But enter Scott and Suzanne Ramsey, who say that they have spent over $500,000 of their own money over a period of 25 years, traveled to 27 states and collected over 55,000 documents investigating the supposed Aztec crash. The result is their 2012 book *The Aztec Incident – Recovery at Hart Canyon*. The Ramseys claim that there are first and (mostly) second-hand witnesses to the crash retrieval operation, and that the honest and successful oilman Silas Newton was pretty much framed by J.P. Cahn to discredit the crash story, and especially discredit Frank Scully. This time, even Stanton Friedman is on board the Aztec saucer: he wrote, "Years ago I was indeed dubious about the Aztec case, thanks to Bill Steinman, Silas

Newton, etc., etc. However, once I did my homework, visited the site, and spent time with Scott Ramsey and others who have been doing theirs, I became convinced that indeed a saucer crashed at Hart Canyon near Aztec in March, 1948, and its retrieval was indeed, as one might expect, covered up by the U.S. Government. Scott has really dug into the documentation, and found witnesses, etc. I would probably still rank Roswell the number one crash." (*Saucer Smear,* May 1, 2006).

One of the government scientists who supposedly studied the crashed saucer was a mysterious "Doctor Gee" (obviously Gebauer despite his and others' denial), who let his friend, oilman Silas Newton, in on the secret as they were driving around trying out a magnetic device for detecting "microwaves" supposedly emitted by oil in the ground. Dr. Gee claimed to be a "master of magnetic energy," and to have worked on a device called a magnetron that "knocked out as many as 17 Japanese submarines in one day." The saucers, according to Newton, probably originate on Venus, and use "magnetic" propulsion, traveling along "Magnetic Lines of Force which originate in the sun and revolve around their planets and their moons, keeping the universe in magnetic balance." The Ramseys don't seem to realize that these statements are pseudoscientific gibberish.

**Silas M. Newton**, according to the Ramseys, has gotten a bum rap from Cahn. His supposedly "questionable" business deals were "nothing new, as the oil business always has a high level of risk… Newton's investors were, in the main, all quite happy, as court records show." (Of course, con men like Newton typically use money from new "investors" to pay off earlier "investors" who threaten to complain to the law.) The FBI, however, takes a different view. Official government records on the **FBI website** contain "the FBI's investigations into Newton's fraudulent activities between 1951 and 1970." According to the FBI, "Silas Newton (1887-1972) was a wealthy oil producer and con-man who claimed that he had a gadget that could detect minerals and oil." Newton's first arrest was as far back as 1931, with many more arrests following. In 1970 Newton pleaded guilty in Los Angeles to illegal securities sales (but was allowed to withdraw that plea after making restitution to the investor), and was also under investigation for an alleged mining fraud in New Mexico, two years before his death at age 85. And the Ramseys state, "the FBI is still withholding over 211 pages concerning Silas Newton." Would you buy a used saucer from that man?

The town of Aztec even provides a map showing would-be crash site investigators how to get there. I started out down that road once, but I was driving an old RV, and had serious doubts about how much of my RV would remain in one piece if I continued eight miles down that horrid dirt road. I understand that there isn't a great deal to be seen there except for a small concrete pad whose origin nobody can explain. (Somehow I can't imagine that a book titled *Small Concrete Pad of the Gods* might become a best-seller, but I could be wrong.) Crash proponents don't like saying how small that "mystery slab" really is, because you'd then realize it would have been useless in any UFO recovery. However, Ramsey lets it slip that the slab is actually only one-meter (39") square, and about 9.5" (24 cm) thick. That is a really small slab to land a flying saucer on, or even to place a crane on to hoist one.

But sunken UFO stories, like Rubber Duckies, have an amazing propensity for popping up to the surface again and again.

## Roswell - The Granddaddy of All UFO Crashes

The one crashed saucer claim to make it big in popular culture was the alleged crash near Roswell, New Mexico, in 1947. However, the event was all but forgotten until it was resurrected by the 1980 book *The Roswell Incident* by Charles Berlitz (of Bermuda Triangle fame) and William L. Moore. You will look in vain for the word "Roswell" in any UFO book or article published before 1980, even if the subject is UFO crashes. A long series of sensationalist movies, TV shows, books, and so on have made Roswell a household name synonymous with UFO aliens. By the late 1990s, it was clear to anyone who cared about facts that the supposed Roswell crash involved a once-secret balloon-borne intelligence-gathering initiative called Project Mogul. The prototype for a balloon-based intelligence-gathering platform that never quite got off the ground, so to speak, Mogul was nonetheless a classified project. This the Army's cover story about a "weather balloon," rather than a "spy balloon," a 1940s version of contemporary drones. The man who actually found the Roswell debris, Mac Brazel, described it as looking like "large number of pieces of paper covered with a foil-like substance and pieced together with small sticks much like a kite" (although the conspiracy theorists insist that this was just a 'cover story' that the military forced him to say!). In other words, it looked like the debris of a balloon's radar reflector, probably from Project Mogul. There is no credible evidence that what was recovered at Roswell was ever anything bigger than tinfoil and sticks. For those who want to read the full details of the Roswell crash investigation, I recommend *Roswell – Inconvenient Facts and the Will to Believe* by Karl T. Pflock (Prometheus Books, 2001). Pflock (1943-2006), who was not a UFO skeptic, started out believing the Roswell crash story, but his careful investigation convinced him otherwise. Once a story of this magnitude plants itself in the popular imagination, nothing will ever lay it to rest. The Roswell legend will live on as long as there are claims of UFOs.

As for the "serious" Roswell evidence: the Center for UFO Studies (CUFOS), which purports to do scientific research on the subject acknowledged in 2002 – with admirable courage - that one of the most famous and credible Roswell witnesses appears to have been lying all along. The late Frank Kaufmann claimed to have served in military intelligence at Roswell in 1947, allegedly working on the retrieval of dead aliens and UFO wreckage. His claims were widely believed by UFOlogists, in spite of certain difficulties that they posed: his name did not appear in the base yearbook for 1947, and some of his statements contradicted other supposed "known facts" about the alleged crash, such as the date. Still, Kaufmann spoke convincingly, had documents to support his tale, and seemed to supply so many credible facts about the "crash" that UFOlogists found it impossible to ignore his account. The pro-Roswell researcher Kevin Randle wrote, "It seemed that every time we began to doubt, or ask difficult questions, Kaufmann would provide another a little bit of documentation along with broad hints that he had much more. He said, repeatedly, when the time came, he had the documents to prove what he said...Kaufmann died in February, 2001, without ever been proved [sic] a liar and a fraud."

The proof, however, was not long in coming. After **Frank Kaufmann**'s death, **CUFOS'** Scientific Director Dr. Mark Rodeghier and two of his colleagues met with Kaufmann's widow to attempt to secure whatever additional documentation he might have had concerning the alleged UFO crash. What they found was truly amazing, beyond what any of them had expected. They found Kaufmann's *true,* undoctored military record that, unlike the copy he showed them, revealed he had never worked in intelligence. Another document, a memo purporting to show Kaufmann's involvement in a "Flying Disk" recovery team, was dated 25 July 1947, signed by Major Lester Garrigues. Rodeghier determined that Garrigues was still living, and got in contact with him. Garrigues not only stated that he had already been transferred from Roswell to an overseas assignment on the date in question, but produced documentation to prove it. Worst of all, it was determined that Kaufmann owned an old typewriter whose type face was an exact match to that on three alleged Roswell documents. Rodeghier concludes, "Given all this evidence of counterfeit documents, we can have no confidence in any details of Kaufmann's testimony."

The problem for the Roswell promoters is, however, that with the credibility of "crash witness" Frank Kaufmann's testimony now having crashed, as previously has that of star "crash witnesses" Jim Ragsdale, Gerald Anderson, and Glen Dennis, little or nothing remains of so-called "crash testimony." In *Roswell – Inconvenient Facts and the Will to Believe*, Karl Pflock demonstrated inconsistencies such that of the just four people publicly identified as witnesses to alien bodies, "not one of the purported firsthand witnesses to alien bodies and a lone survivor is credible. Not one." (Pflock, 2001, p. 118-120). Those who cite supposed 'crash dummies' parachuted into the desert as an explanation for Roswell testimony about 'little bodies' do not seem to realize that there is no such testimony left unrefuted.

We are now also hearing about a supposed 'deathbed confession' of **Lt. Walter Haut** (1922-2005), the public information officer at Roswell who on July 8, 1947 drafted the now-famous press release stating that the army had recovered a crashed "flying disk" on a ranch near Roswell. Pro-Roswell researcher Donald Schmidt drafted an **affidavit** in 2002 for the then-infirm Haut to sign, stating that he had personally seen a crashed saucer and little bodies at the Roswell base, something Haut had never before claimed. But another pro-Roswell researcher, **Kevin Randle**, notes that in other interviews given around that time, Haut gives "a rambling mishmash of contradictory information." Randle suggests, quite sensibly, that by the time Haut had reached approximately age eighty, his mental state was sufficiently confused that we cannot rely on anything he then says.

The *Science Fiction Channel* (now *SyFy*) presented a supposed documentary *The Roswell Crash – Startling New Evidence*, which premiered on Nov. 22, 2002. Prior to the broadcast, network representatives were promising it would contain a "Smoking Gun," although after seeing the show the only thing smoking seemed to have been the show's producers. Shovels in hand at the supposed Roswell "crash site," (actually, one of several alleged sites), the researchers had found *something,* they tantalizingly hinted. That *something* turned out to be nothing more than evidence of a furrow in the ground, which might have been caused by an alien crash – or by a burrowing animal. In the most shameless example of TV hucksterism since Geraldo Rivera's prime-time excavation of what he called "Al Capone's vault" uncovered precisely nothing, *The Roswell Crash* ended by ostentatiously carrying *bags of dirt* under heavy armed guard to be placed in a bank vault, because nobody had proven that they did not contain microscopic extraterrestrial artifacts.

But the "smoking gun" was actually nothing more than researcher David Rudiak's earlier claim to be able to read about "victims" of a saucer crash in a blurry news photo of General Ramey holding a memo while reciting the supposed Roswell "cover story" about a balloon. Rudiak claims that while General Ramey was talking to the press, he was holding in his hand a top-secret memo containing the phrases like "victims of the wreck" and "in the disc." Of course, as everyone knows who has ever worked on classified military projects, it violates every rule for handling classified documents to take one out of a secure facility into the open where it might be seen by uncleared persons – especially to someplace as public as a press conference! Had anyone actually done this, he would have lost his security clearance at once. Rudiak states,

> There is no question that Ramey's message, even when greatly enlarged and then enhanced by computer, is a very difficult read because of fuzziness, film grain noise, uneven development, photo defects, paper folds and tilt, shadows, and text obscured at the left margin by Gen. Ramey's thumb. This will inevitably prompt comments from die-hard skeptics that my full "take" on the Ramey message is strictly in my imagination

He is right about that!

Yet another bizarre "explanation" for the Roswell story has surfaced, in a book *Body Snatchers in the Desert: The Horrible Truth at the Heart of the Roswell Story* by the British UFOlogist Nick Redfern, who went on Coast to Coast AM on June 21, 2005 to tell his tale. According to Redfern, what happened in the new Mexico desert in 1947 was not a flying saucer crash, but something much worse: "A highly confidential, U.S. government-sanctioned program to conduct medical experiments on deformed, handicapped, disfigured, and diseased Japanese POWs, exploited as 'expendable' victims by their captors." In an interview, Redfern explains, "when the war was coming to a close, there was a particular skirmish on an island in the Pacific between the Americans and the Japanese. There was a scientific medical laboratory there which was allied to the Japanese government's notorious unit 731. Now unit 731 was literally the equivalent of what the Nazis were doing.... they also brought back survivors from the Unit 731 camps. People who had been experimented on, who were due to be experimented on, even a number of dead bodies." Allegedly the U.S. military continued these inhumane experiments, and also used some of these unfortunates to test balloons and "really advanced aircraft and gliders." When one or more of these crashed in the desert, witnesses saw the emaciated, radiation-shriveled bodies of the small-stature Japanese prisoners, and concluded that they had seen space aliens.

However not everybody was climbing on board Redfern's bandwagon. Stanton Friedman quite correctly cites problems with Redfern's "history", and with his anonymous sources. Karl Pflock noted many more problems, concluding, "With the original stories [of Roswell "alien body witnesses"] discredited, there is no foundation for what you recount."

## Commie Nazi Saucer Crashed in Roswell

Just when you think that the credibility of the mass media couldn't go any lower, they fell for this one. According to a 2011 book by Annie Jacobsen, *Area 51: An Uncensored History of America's Top Secret Military Base*, what crashed at Roswell was not an alien saucer, but instead a collaboration between the communist Stalin and ex-Nazi Mengele to cause panic in the U.S.

This absurd claim was presented in all seriousness by many of the major media, including the *Huffington Post*. The *Albuquerque Journal* (May 22) headlined, "Stalin, Nazi Implicated in Roswell Incident." The story was also reported in the *Boston Globe*, *AP News*, etc. Practically the only news outlet to show any proper skepticism was Bloomberg's review of the book, "Roswell Martians Might Have Been Nazi Kids From Mengele's Lab."

# Bad UFOs

No reporter worth his salt should ever have given credence to this preposterous claim, even for a minute. Any one who did belongs in some other profession. There are so many things wrong with this claim, one hardly knows where to start. But I'll give it a try:

The first publicized sighting of a disc-shaped craft was that of Kenneth Arnold on June 24, 1947. The Roswell incident occurred *ten days later*. Did Stalin and Mengele put this hoax together in just ten days? Jacobsen speculated that the 1938 Orson Wells *War of the Worlds* panic gave Stalin the inspiration for this stunt. And he did all this while fighting the Germans during World War II?

How did this thing fly? Its propulsion system would be immediately recognized by U.S. experts as conventional and terrestrial. Its electronics would also be immediately recognized as Soviet in origin, years behind the best U.S. made systems. The idea that Stalin could fool U.S. experts into thinking that they were examining an extraterrestrial craft is simply absurd. Stalin was indeed devious and evil, but he was not stupid.

The children who had allegedly been deformed by the cruel Dr. Mengele to look like extraterrestrials reportedly did not all die in the crash. Surely their Soviet handlers knew that if any of them survived, they would immediately be screaming for help in Russian or German, and would reveal the deception at once.

But remember, like the supposed parachuting crash dummies, this is an attempt to explain something that does not need explaining. Jacobsen's wild claim attempts to explain the origin of accounts of a disc-shaped craft that allegedly crashed at Roswell, containing alien-looking beings. But there is no credible evidence whatsoever that anyone actually saw this. It is not wise to invoke extraordinary explanations for ordinary phenomena. This is the Fallacy of Misplaced Rationalism, when the "rational" explanation is even more implausible that what it is supposed to explain (see my "Psychic Vibrations" column in the Skeptical Inquirer, July/August, 2008. Reprinted in *Science Under Siege*, Kendrick Frazier, ed.).

As anyone who has read Solzhenitsyn's *Gulag Archipelago* surely realizes, Stalin never gave anyone who might have information embarrassing to the Soviet state a chance to tell it. For example, at the close of World War II millions of returning Soviet prisoners-of-war were sent to the Gulag for the "crime" of having first-hand knowledge of the outside world. We know that Mengele was hiding out in Bavaria under an assumed name from 1945 to 1949, and escaped to Argentina soon afterward - all while Stalin was still living. Thus the known facts of Mengele's life contradict this preposterous tale - at no time was he ever living in the USSR, or even in the USSR-occupied sector of Germany. If Stalin ever held Mengele, he would have placed him in the Gulag, probably in the "First Circle" (see Solzhenitsyn), where scientific and technical prisoners were held, but compelled to work for the Soviet state. Stalin never would have let him escape, and Stalin lived until 1953, by which time Mengele had already been in Argentina for several years. But when did the major media ever let facts get in the way of a sensational story?

## The Hottel Memo: FBI Admits UFO Crash at Roswell?

In April, 2011, the major media made such a big thing out of a 1950 memo written by FBI Agent Guy Hottel, repeating a yarn he'd been told. For example, see this piece from *The International Business Times* (April 10, 2011, and now acknowledged as a hoax):

> A secret memo released online by the Federal Bureau of Investigations (FBI) in its 'Vault' has emerged as proof for the famed landing -- or crash or capture -- of a flying saucer with three dead aliens in Roswell in New Mexico in June 1947.

> The memo, titled 'Flying Saucers', was written by FBI agent Guy Hottel. The decades-old memo, which was published by the FBI in its 'Vault,' says "three so-called flying saucers had been recovered in New Mexico', citing an Air Force investigator.

Ben Radford pointed out in his *Livescience* blog that the FBI Memo was far from new, and that it had nothing to do with Roswell. Dave Thomas of the **New Mexicans for Science and Reason** has written an analysis showing that William L. Moore had already explained the origin of the claims in the Hottel memo to the 1985 MUFON Symposium – and found it traces back, eighth-hand, to *claims made by con-man Silas Newton about the supposed Aztec crash*!

one of the agents at OSI headquarters in Washington, passed the Fick story, now seventh-hand, along to Special Agent Guy Hottel, one of his contacts in the Washington office of the FBI (with whom OSI often worked quite closely), who in turn, on March 22, 1950, generated a memo on it to J.Edgar Hoover himself. Hottel's memo, repeating a now eighth-hand story but still retaining the four key points of the original Newton story (i.e. high-powered radar site in New Mexico [but now without mention of Arizona], three foot tall aliens, metallic cloth and wrapped bodies), has been cited out of context again and again by an entire array of UFO researchers as conclusive evidence that the U.S. Government is in possession of a crashed saucer. Had any of them bothered to research the matter before jumping to conclusions, they would have realised the memo is essentially useless in that the origin of the information cited therein can be traced directly to Silas M. Newton himself.

In my files I found a 1983 newspaper article showing how MUFON's Director, Walt Andrus, was already shamelessly trying to exploit the **Hottel Memo.**

The lesson of this whole fiasco is: the next time you hear the major news media reporting some major development supposedly proving the reality of a government cover-up of UFOs, remember how they sensationalized the sixty-year-old, eighth-hand information in the FBI memo. And remember that they're in the business of grabbing readers first, and only secondarily of reporting the news in factual and accurate manner.

## Alien Ghosts at Roswell?

Is it possible that, once in a great while, a pseudo-scientific documentary is so insulting to the viewer's intelligence that even the cable channel funding it turns it down? *The Alien Ghosts at Roswell* was produced by Jim Marrs, a well-known author and conspiracy theorist. Marrs first became famous making JFK Assassination Conspiracy claims, then moved on to "the extraterrestrial presence," psychic spies, secret societies, and 9-11 conspiracies. The documentary was made for *The Discovery Channel*, which has not shied away from many other programs making absurd claims about hauntings, life after death, numerology, etc.

For several years, Marrs has been promoting the story that not only did a saucer crash near Roswell in 1947, but some of the alien beings killed in the crash left behind ghosts. The former air base hospital, now the New Mexico Rehabilitation Center, supposedly contains one wing where "none of the personnel want to work." The reason for this is said to be "because of strange and unexplained incidents which have occurred there." Lights are said to be turning themselves off and on, the elevator doors would unaccountably open and close, and areas of intense cold are reportedly felt. A strange figure has reportedly been appearing in the upstairs hallway. "It all seems to take place on the second floor near the room where they supposedly did an autopsy on the aliens," said systems analyst David Owen. "I've heard people say they have seen aliens running around."

To exploit his remarkable "scoop," Marrs produced a 45-minute documentary titled *The Alien Ghosts at Roswell*. However, after the folks at the Discovery Channel saw the video, they didn't want it. Fortunately, UFOlogist James Moseley was able to get a copy, which he reviewed in his publication *Saucer Smear* (Oct. 20, 2007). According to Moseley, the video, which was filmed at the old Roswell air base, and "stars a group of psychic researchers, led by ufologist Nick Redfern." One scene shows an animated alien ghost walking down the hallway, to illustrate what has supposedly been seen.

Moseley wrote, "we are told that [the Discovery Channel] paid off Jim Marrs' entire contract and went on to other things. Thus it will probably never be seen by the public, alas." So far that seems to be the case, although Marrs continues to talk about alien ghosts on the paranormal radio show *Coast to Coast AM*, and other places where silliness is the norm.

## The Fiasco of the Roswell Slides

Sometime during 2012, the Chicago-based video producer Adam Dew obtained a collection of Kodachrome slides reportedly taken during the 1940s. The slides were said to have been taken by the late Bernard and Hilda Ray, a well-to-do Texas couple who led an active life with much travel, and left behind no family. Two of the slides were of particular interest: they seemed to show the body of a small being laid out on a shelf. It looked like it might be an alien, Dew thought. So he contacted

'Roswell experts' Donald Schmitt and Tom Carey, authors *of Witness to Roswell*. At that time, those authors were involved putting together something to be called the "Roswell dream team," intended to bring together expert investigators to do a fresh evaluation of the Roswell incident, and hopefully obtain proof that the crashed saucer story was real.

UFO investigator Anthony Bragalia was a member of this team, as were Kevin Randle, a big Roswell proponent and author of many UFO books, and the Canadian investigator and author Chris Rutkowski. However, the harmonious Roswell dreaming was soon interrupted. Randle and Schmitt had been partners in earlier Roswell investigations during the 1990s. However, when it was discovered that Schmitt had falsified his credentials, among other things, Randle denounced him, and severed all cooperation. After about twenty years, they were beginning to reconcile, when the slides turned up. Randle stuck his toe into the water (or perhaps his whole foot), but didn't like what he saw, and withdrew from the effort. (In early 2015, as the Slides fiasco gathered intensity and momentum, Randle sent me an email essentially saying, "I hope you realize that I have nothing at all to do with these Roswell slides!" I assured him that I did.) Rutkowski says that he was approached about being a member of the Dream Team, but when he expressed some reservations about the slides, his "membership" offer was withdrawn.

The remaining members of what should now be called the Slides Team apparently had no reservations whatsoever. The slides were supposedly being investigated by the best photographic and other experts, who said they appeared to be authentic. The cardboard mountings of the slides were said to prove that they must have been processed during the 1940s. The only problem was, nobody outside that group had actually seen the slides, and the details of the supposed investigations were hazy. We were assured that when the time was right, and the investigations were complete, this "smoking gun" evidence of the Roswell crash would be released to the world.

For about two years, the existence of the slides was known mostly just to those who follow UFO-related blogs and such, and their content was only rumored. As might be expected, curiosity about them was building, along with a properly skeptical "wait and see" attitude. Then at a public forum in November, 2014, Tom Carey announced,

We have come into possession of a couple of Kodachrome color slides of an alien being lying in a glass case. What's interesting is, the film is dated 1947. We took it to the official historian of Kodak up in Rochester, New York, and he did his due diligence on it, and he said yes, this filmstrip, the slides are from 1947. It's 1947 stock. And from the emulsions on the image, it's not something that's been Photoshopped like today. It's original 1947 images, and it shows an alien who's been partially dissected lying in a case.

He described the being as "3 and a half to 4 feet tall, the head is almost insect-like. The head has been severed, and there's been a partial autopsy; the innards have been removed, and we believe the cadaver has been embalmed, at least at the time this picture was taken. The owners of the slide -- it's an amazing story. The woman was a high-powered Midland, Texas, lawyer with a pilot's license. We think she was involved in intelligence in World War II, and her husband was a field geologist for an oil company." This was reported in the press, and noted widely.

Dew released on YouTube a professionally-produced teaser for his in-production documentary motion picture titled *Kodachrome* to hype the slides, from which blurry images of the two slides leaked, perhaps intentionally. A loosely-organized group of independent investigators came together, calling itself the Roswell Slides Research Group. But it is difficult to investigate something that one is not allowed to see, except as a small icon. Anthony Bragalia wrote on **Before Its News** (July 14, 2014),

> Of the many scientists, PhDs, photography experts and other researchers who are among the very fortunate to have viewed the 'alien slides'- not one has ever at any time mentioned that the 3 foot thing depicted in the 1947 photographs resembles a mummy. This includes KODAK experts, a NASA scientist of international standing who has left comments on his impressions of the creature on this blog and several UFOlogists. The creature depicted in the slides (owned by an Oil Exploration Geologist in NM in the 1940s) in no way even remotely appears like any creature known on Earth.

Surprisingly, the "alien" body seems to have a placard on the shelf next to it. Why would a top-secret dead alien hidden away in Area 51 sport a placard, like a mummy in a museum exhibit? What would such a placard say? "Dead alien from Roswell. Top Secret – Don't Tell Anyone!" Was it possible to read the writing on the placard? Tom Carey said "there's a placard, very fuzzy, that can not be legibly read by the naked eye, yet we've had everyone from Dr. David Rudiak, to Studio MacBeth, even the Photo Interpretation Department of the Pentagon, as well as Adobe have all told us that it's beyond the pale, that it cannot be read, it is totally up to interpretation."

A date was finally set to reveal the slides: May 5, 2015 – in Mexico City. Mexico City? Yes, indeed. A big extravaganza was being planned to reveal the slides on *el Cinco de Mayo*, and it was organized by Mexico's best-known UFO huckster, Jaime Maussan. If the slide promoters were seeking credibility (as opposed to a quick buck), they could not have made a worse choice. A well-known sensationalist journalist, Maussan is Mexico's very own P.T. Barnum, having made a lucrative career peddling dodgy photos and videos of UFOs, alien beings, and the like. He previously promoted a skinned dead squirrel monkey as an alien creature, and even published a photo of what is supposed to be "un caballo en el cielo" – a horse flying across the sky. Alejandro Franz's website *alcione.org* lists "more than 40 frauds of a pseudo-journalist and charlatan," Maussan.

Maussan hyped the Slides shamelessly, promising to reveal a Roswell "Smoking Gun" on May 5. This was your opportunity to witness an event that would "change history!" About 6,000 tickets were sold priced between $20 and $100 US (according to Ticketmaster in Mexico); some accounts claim that up to $350 per ticket was paid. Thousands paid $15 or $20 to watch the bi-lingual event on streaming internet video (which did not work well, angering many). The Twitter feed of those watching the streaming video (#RoswellSlides) was overwhelmingly negative, with most commenters mocking the presentation.

Promises to release high-resolution copies of the slides after the presentation did not materialize. However, at least one "high enough" resolution copy of the slide showing the placard did leak out. The French skeptic Nab Lator of the Roswell Slides Research Group quickly used the commercial software Smart Deblur to read the placard. The first line clearly read, "MUMMIFIED BODY OF TWO YEAR OLD BOY." Others quickly confirmed that finding. Researcher and satellite orbit guru Ted Molczan commented, "You folks solved in no more than 2-3 days what the promoters claimed not to have been able to solve in 3 years!" Perhaps the slide promoters had an incentive *not* to be able to read the placard.

When the deblurred copy of the placard was released on the internet, Dew was furious. He called the Roswell Slides Research Group "a group of internet UFO trolls, claiming to be searching for truth but repeatedly spreading lies." He claimed that they created a "fake placard" using Photoshop, and hastily posted a copy of the 'authentic' blurred placard on his own website, removing all doubt as to the provenance of the placard image. However, instructions were soon posted by slide debunkers showing how to take the copy of the placard *from Dew's own website*, and de-blur it to read at least the top line in less than two minutes using a trial copy of Smart Deblur. As for Bragalia, he quickly issued a somewhat tepid *mea culpa*, saying "I must be less trusting, more discerning and less accusatory of those with whom I disagree," and he blamed Dew for withholding the high-resolution scans from independent analysis. Considering that Bragalia had earlier called the detractors of the yet-unseen slides "rabid slide-skeptics" and even worse, some think that he has a lot more apologizing to do.

Amazingly, in the same posting in which Bragalia switched sides on this, he claims to have ID'd the specific mummy trumpeted as the 'Roswell alien': it is a child mummy from Native American cliff dwellings in Arizona, at one time on display in a museum in Mesa Verde National Park in Colorado. About a week later, Donald Schmitt seemed to throw in the towel by saying he had been "overly trusting," but then later made other statements suggesting that the slides might still be genuine. Tom Carey said that the matter "is still open to debate." As of this writing, there is no *mea culpa* from slide promoters Maussan, Dew, and Richard Dolan, with Maussan still insisting that the slides show an alien, joined by other UFO delusionists like Linda Moulton Howe and Whitley Strieber. The most interesting rationalization thus far comes from Dr. Richard O'Connor, M.D., who acknowledges that the placard does indeed read 'MUMMIFIED BODY OF TWO YEAR OLD BOY," but it is a deliberate deception.

So while the fiasco of the Roswell Slides was a huge embarrassment to 'Roswell research,' and to UFOlogy in general, there were nonetheless some hopeful aspects of it. Many well-known pro-UFO researchers were very skeptical of claimed "smoking gun" photos of unknown origin and content, including Stanton Friedman, Kevin Randle, and Nick Pope. More encouraging still was the excellent cooperation between skeptics and UFO proponents, instead of the usual acrimony. Skeptics like Tim Printy, Nab Lator, Gilles Fernandez, and Lance Moody worked alongside open-minded UFO proponents like Paul Kimball, Curt Collins, Isaac Koi, and Chris Rutkowski to cooperatively solve the riddle of the Roswell Slides.

## 1965: Extraterrestrial Acorn Crashes in Kecksburg, Pennsylvania

But one alleged **UFO crash** that's currently getting a lot of attention took place a long ways from Roswell: in **Kecksburg, Pennsylvania**. According to witnesses, just before 5PM on the clear winter afternoon of December 9, 1965, a "fireball" that "seemed to be under some type of intelligent control" appeared to crash in the woods near Kecksburg. State police responded to reports of an object crashing in the woods, but didn't find anything. But according to some accounts, the military cordoned off the area, and hauled off an acorn-shaped metal object to some unknown destination.

The appropriately-named Science Fiction channel (now renamed *SyFy*) premiered a sensationalist pseudo-documentary on October 24, 2003 titled *The New Roswell: Kecksburg Exposed*. As one might imagine, it leaned heavily toward the claims that a saucer crashed and was covered up by the military, giving short shrift to any skeptical explanations.

The only problem with the so-called Kecksburg Crash is that the object was identified long ago, was reported to have crashed in many different places, and is known to have actually disintegrated above Ontario. What the UFO proponents are calling the "Kecksburg crash" is known to astronomers as "the Great Lakes Fireball of Dec. 9, 1965." Indeed, it is one of the best-studied fireballs in history, because of the clear skies and mild weather that permitted large numbers of people to see it. It has been written up in leading astronomy journals, including *Sky and Telescope* magazine and the *Journal of the Royal Astronomical Society of Canada* (for the details see my ***debunker.com*** web page on **Kecksburg**). Indeed, so much information was available on this brilliant meteor that astronomers were able to determine the orbit it was in *before* it encountered the earth. It was in an eccentric orbit with a period of approximately 2.43 years, taking it out past the orbit of Mars at its farthest point from the sun, to just inside the earth's orbit, where the rock met its doom.

Even before any UFO claims began to surface about it, the Great Lakes Fireball was cited as an illustration of the unreliability of eyewitness testimony. Prof. G. W. Wetherill, a professor of geophysics and geology at UCLA who investigated the incident, was quoted in *Sky and Telescope* magazine (February 1966):

> The fireball was observed by many people in Ontario, Michigan, Ohio, Pennsylvania, and to a lesser extent in neighboring states. In newspaper accounts, a great many supposed impact sites were reported, both in southwestern Pennsylvania and eastern Ohio. Fragments were claimed to have fallen in Ohio and Michigan. These imagined happenings arose from the impossibility of estimating the distance of an object in the sky. Almost everyone who saw the fireball thought it was much closer than it really was. When it disappeared behind a house or a tree many people thought it had fallen only a few hundred yards beyond.

Nor is any of this information particularly new. Astronomer Robert Young debunked the "Kecksburg crash" claims after they first appeared on the TV show *Unsolved Mysteries*" in 1990, which got extremely high ratings (see "Old-Solved Mysteries: The Kecksburg Incident," *Skeptical Inquirer*, Spring, 1991). Young's paper was revised and reprinted in the book *The UFO Invasion* (Frazier, Karr, and Nickel, eds. Prometheus Books, 1997). It has been carefully ignored by UFO proponents.

The *Science Fiction Channel* has clearly done an extremely good job of leading the public astray. A poll on its website revealed that 67% of those taking the poll believed that the object at Kecksburg was an "alien craft," versus only 27% who chose one of several prosaic explanations. But the next time you hear someone touting Kecksburg as "the new Roswell," at least you'll know that the new crash claim is just as bogus as the old Roswell.

## Crashes Here, Crashes There, Saucer Crashes Everywhere

Today, the list of alleged UFO crashes has expanded far beyond the few familiar names like Roswell, Kecksburg, and Aztec. Claims about UFO crashes and their cover-ups make up a major part of contemporary UFOlogy. For several years starting in 2003 a "Crash Retrieval Conference" was held each November in Las Vegas. It was organized by Ryan S. Wood, who claims that there have been at least seventy-four UFO crashes worldwide.

The UFO group **CSETI** lists on its web page 272 possible **UFO crashes** worldwide since prehistoric times (not including those on the Moon, or on Mars). In the US alone, saucers have allegedly fallen from the skies near Kingman, Arizona; Las Vegas, Nevada; Laredo, Texas; Death Valley, California; and Paradise Valley, Arizona, to name just a few, not to mention a second crash at Roswell two years after the famous one. For unknown reasons, there seems to be proportionately fewer UFO crash reports from the populous eastern states. Perhaps the aliens should be wary of the deadly Arid Southwest Triangle with a mysterious force in that pulls saucers earthward.

When I was attending the 2011 MUFON International Symposium in Irvine, California, I was standing not far from the famous "UFO abductee" Travis Walton's table where he was selling books, illustrations, etc. One fellow, tall, thirtyish, dressed in a nice suit, came up to tell everyone how he had evidence of a UFO crash in Idaho, with lots of debris scattered over a large area. He didn't just see the debris, he witnessed the UFO crash. He had photos, all kinds of proof. Can we see it? "The photos are on a flash drive. I left it in the other hotel when I came here in the taxi." At first his story was that all that was needed was to go back in a taxi to get the photos and bring them here. But soon he was objecting that he was going to be writing a book about it, and he didn't want anyone to steal his discovery. The objections came swifter and stronger. It felt strange for me to be joining Travis Walton in critiquing a preposterous claim, but that's what happened. Apparently Mr. Idaho Crash told his story to a lot of people there, but offered absolutely no proof. I asked myself, "Why does this man crave attention so desperately to keep telling a story like that?"

MUFON tried to get people to attend its 2012 International Symposium held in northern Kentucky near Cincinnati, Ohio, by promising a "blockbuster UFO discovery." The Symposium schedule for Sunday read,

Sunday August 5

9:00am-4:00pm: National Air Force Museum at Wright–Patterson Air Force Base

4:30pm-6:30pm: National Release of "Blockbuster" UFO discovery

And what exactly was the "Blockbuster" UFO discovery? As **Jack Brewer** wrote in **The Examiner**,

> researcher **Harry Drew** took the podium in Covington, Ky., explaining he is convinced he has located two sites where alien craft landed or crashed in 1953. The sites are in the vicinity of Kingman, Ariz. Drew explained he believes the craft were brought down by triangulated radar running at boosted power to extend range. According to Drew, military personnel quickly retrieved and cleaned up the wreckage. The archaeologist and historian apparently included photos of the sites in his presentation, as well as presented information contradicting past accounts of the case and alleged crashes.

Wow! More stories (without proof) about saucers that supposedly crashed sixty years earlier! It's a Blockbuster! (Actually, it's very Old News: a Google search of **UFO CRASH KINGMAN 1953** returns 160,000 hits.) Brewer adds, "Don't shoot me, I'm just the piano player."

The Blogosphere was quick to dismiss MUFON's grandstanding. The UFO and Conspiracy website *GodlikeProductions* quickly sported a thread, "MUFON 'Blockbuster' Announcement Peters Out Without A Bang." The *Following the Nerd* "Paranormal" Blog said, "The big build up to an earth shattering announcement from MUFON yesterday that they would rock the UFO community to their core proved to be more than a disappointment."

## 1974: Carbondale, Pennsylvania Flashlight UFO Crash Coverup

Perhaps sensing a profitable venture, some folks are working hard to rehabilitate the once-infamous (and now virtually forgotten) alleged UFO crash at Carbondale, Pennsylvania. On the night of November 9, 1974, three teenage boys reportedly heard a noise in the sky, then looked up to see a red ball of light coming at them from over a nearby mountain. They claimed it plunged into a small pond, from whose depths a light could be seen glowing faintly. The authorities were notified, and thus began a two-day vigil next at the water's edge. Finally, after Geiger counter readings showed that the pond was not radioactive, a determined diver named

Mark Stamey braved the depths and soon returned with an amazing find: a battery-powered railroad lantern. It was never determined who had thrown it into the pond. I wrote about this in *UFO Sightings* (Sheaffer, 1998, Chapter 4).

Aah, but more than thirty years later, "the truth" is coming out: According to BUFO, **Burlington UFO Internet Radio** (February, 2006), "the 'lantern story' was a coverup for something much bigger - something that our government did not want us to know about." Diver Stamey is now claiming that the lantern incident was "staged," and a UFO remained in the water's depths. "He claims that the lantern was seeded there by acting Sheriff .. He further claims that the Sheriff ordered him to go into the waters and pull this lantern out for the purpose of thowing off the public and media as to what was really moving around down there...a ufo!"

## Long Island UFO Crash Coverup: The Radioactive Toothpaste Revenge

On November 24, 1992 at 7:00pm, people reportedly saw something crashing into the woods of South Haven Park, on Long Island, New York, near the Brookhaven National Laboratory. Supposedly the area was cordoned off for four days, with "black suited SWAT teams blocking all access to the area." Supposedly what crashed was a "mothership" with "six escape pods," according to UFOlogist Stephen L. Schwartz.

A UFOlogist promoting this claim, **John Ford** of the **Long Island UFO Network** (LIUFON), felt that he had solid evidence for the crash, but that local officials were obstructing him. Ford was arrested in 1996 in a bizarre plot to attempt to murder local government officials who he held responsible for the alleged crash cover-up. The plan was to poison them by putting radium in their food, their cars, and even in their toothpaste. Mr. Ford was found mentally unfit to stand trial, and was sent to a mental institution. However, some UFOlogists still insist that Ford was framed by cover-up agents, and that his arrest, intended to intimidate those who might pry into UFO secrets, illustrates the depths to which the authorities will go to keep the subject under wraps. There is a **John Ford Defense Committee** working to get him sprung. According to that committee,

## UFO Crashes and Retrievals

On June 13, 1996, the most prominent UFO investigator on Long Island, John Ford, was arrested on the absolutely ludicrous charge of conspiring to murder John Powell, the Suffolk County Executive and head of the local Republican Party, by conceiving of a plan to put radium in his toothpaste. The only evidence they had was a tape recorded conversation that was altered and proof of that alteration is now available. Police raided his house where they confiscated his UFO files and found non-lethal radium. Although they procured a search warrant, it was dated June 14th and after the arrest had taken place.

As of this writing, Ford remains in the Mid-Hudson Forensic Center for the criminally insane.

A more recent example of a crash hoax was the **You Tube** video of a supposed glowing-hot "**meteorite**" crashing in **Latvia** in October, 2009, belching flame and smoke. While initial news reports were supportive of the claim, experts soon began noticing problems. The bad acting in the video as the students "discover" the meteorite did not help things. The supposed "crater" was seen to have been dug by shovels, and it was pointed out that by the time meteors land they're no longer hot enough to glow. Still, the hoax kept people guessing for a while.

# 5 THE RISE AND FALL OF UFO ABDUCTIONS

Americans seem obsessed with the idea of people being abducted by UFOs. And that abduction typically has overtones of sex. I single out "Americans" because no other country seems to have embraced the idea as fully. In 1996 when I was on a tour of the UFO hotspots in Mexico (see chapter 21 of my *UFO Sightings (Sheaffer 1998)*), the locals told us all kinds of implausible stories about UFO sightings and alien visitors (although we didn't see *anything* ourselves). But then, after telling us about all of the alien craft that he or she had sighted, the Mexican UFO enthusiast would often add "But those abduction cases you people have in the U.S. – I just can't accept that at all!"

Many factors clearly denote the subjective nature of the entire "UFO abduction" phenomenon. The appearance and behavior of supposed UFO occupants varies greatly with location and year. UFO-abduction claims were made much less frequently outside North America, especially in non-English-speaking countries, although foreign reports started to catch up after the overseas publication of Whitley Strieber's *Communion*. The descriptions of supposed UFO aliens contain clear cultural dependencies; in North America large-headed gray aliens predominate, while in Britain abducting aliens have been mostly tall, blond, and Nordic, while South Americans tend to be abducted by more bizarre creatures, including hairy monsters. If we are to believe that such reports reflect reality, then the Galactic High Command must have divided the earth into Alien Occupation Zones whose boundaries reflect those of human culture. Also, the demeanor and activities of supposed UFO occupants depend greatly on which abductionist "uncovers" the story. Each abductionist has a unique "agenda" representing his or her own expectations concerning a UFO abduction, and each case investigated has a tendency to confirm that "agenda." Planting his tongue firmly in cheek, Philip J. Klass gives the following advice to those who think they may have been abducted:

If you are considering contacting an abductionist to undergo regressive hypnosis,... I strongly recommend Dr. Leo Sprinkle. Under his guidance you are more likely to encounter a gentle and spiritually uplifting type of UFOnaut than one who takes flesh samples, ova, or sperm, or removes unborn children from their mothers' wombs (Klass, 1988: 194).

We can find all the major elements of contemporary UFO abductions in a 1930 comic book adventure, *Buck Rogers in the 25th Century*. In it, a scantily-clad Wilma is grabbed by a giant metal clamp coming down from a circular craft and hauled aboard. She calls out in vain for Buck to come rescue her. She is confronted by "Tiger Men from Mars." She is placed on a circular examining table, in a panel where her long and very sexy bare legs are on prominent display. [Dille, 1969, p. 142-5] UFO abductions had a lot to do with sex in 1930, as they do today.

*UFO Abduction in Buck Rogers comic, 1930*

Wilma undergoes a mind-scan. She confers with a subordinate alien, then with the leader. She experiences "theophany" while gazing in awe at the earth from space. This sounds exactly like any late 20th century abduction narrative from a well-known abductionist. UFO investigator and folklorist **Eddie Bullard** has compiled a list of '**elements**' of UFO **abductions** that, if they occur, and in a certain order, are said to be evidence that an alien abduction is 'authentic.' Kottmeyer notes the parallel between reported alien encounters and the elements of drama, suggesting that Bullard's "elements" of abduction [Bullard, 1987, p.47-53] can be understood as what is requisite for good psychodrama. Only one account in Bullard's catalog has a greater number of correctly-ordered abduction elements than does Wilma's in the *Buck Rogers* comic [Kottmeyer, 1990].

*Wilma undergoes a table examination by alien beings, 1930.*

Aliens that steal sperm, eggs, and fetuses, or make scars or body implants on those supposedly abducted, were practically unknown before the publication of Budd Hopkins' books. This particularly alarming type of abduction seems to be quite rare outside North America.

The very first publicized claim of UFO abduction came from Brazil. Farmer Antonio Villas Boas was plowing his fields on the evening of October 16, 1957 when he said that he saw a "red star" appear. As it came in closer and landed, he could see that it was a circular craft. He says it landed, and he was abducted by UFO creatures and then seduced by an alluring, nude extraterrestrial female. The case produced more snickers than genuine interest, although it is sometimes given more respect today given UFOlogy's inexorable trend of steadily increasing credulity.

## Mr. And Mrs. Hills' Wild Ride

Something genuinely new under the UFO sun occurred in 1966: the publication of *The Interrupted Journey* by John G. Fuller, a book detailing the alleged UFO abduction in rural New Hampshire of Betty and Barney Hill. The book reads like a thriller, telling the tale of an interracial married couple driving rural roads late at night, seemingly pursued by a UFO. Upon returning home, Betty began having nightmares about being abducted by aliens. The Hills belatedly concluded that there was "missing time" and came to believe the abduction dreams may have reflected reality. Barney was under much stress prior to the "abduction" and got worse afterward, so the Hills sought therapy from a well-known

# UFO Abductions

psychiatrist, Dr. Benjamin Simon. Under hypnosis, they each told a UFO abduction story that largely matched Betty's nightmares (which Barney had heard her repeat many times).

The Hills' story became a sensation, serialized in *Look* magazine, and was made into a TV movie, *The UFO Incident*. Soon others began making similar abduction claims. The famous Travis Walton abduction story, depicted in the movie *Fire in the Sky*, occurred just a few weeks after *The UFO Incident* aired. Typically, these abduction stories followed a general pattern: You are driving in a rural area at night. You see a light in the sky that seems to be coming closer. You become frightened, and you are unable to recall exactly what happened next. A UFO researcher helpfully puts you under hypnosis, and you suddenly recover repressed memories of an alien abduction. The paradigm of the Hills' abduction prevailed during the 1970s. Persons out on lonely roads late at night risked, in addition to usual earthly perils, abduction by extraterrestrials.

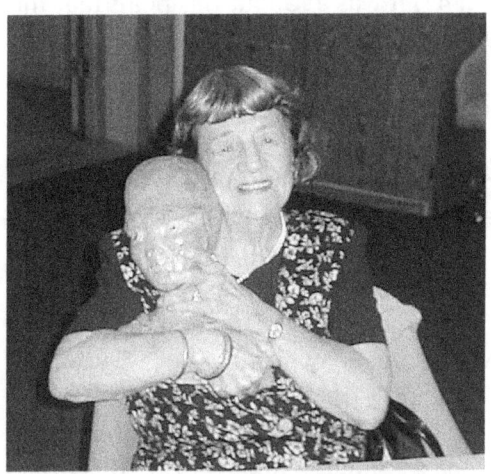

*The author's photo of Betty Hill holding "Junior," a bust of her supposed alien abductor.*

Barney and Betty had been taking a brief and rather hurried vacation trip from their home in Portsmouth, New Hampshire. They went to Niagara Falls and into Canada, returning late at night on September 19-20, 1961. Between 10:00 P.M. and midnight, the Hills were driving south along Route 3, nearing New Hampshire's White Mountains. They report that the sky was perfectly clear. A gibbous moon (more than half full) was shining brightly low in the southwest. As they passed the town of Lancaster, Betty Hill reported seeing a "star" or planet below the moon. Soon afterward, she reportedly spotted a second object, which she described as a bigger or brighter star, above the first object (Fuller, 1966: 171). This was the object that she believed to be a UFO.

They watched this object, or "craft," for at least thirty minutes. It appeared to be following their car. Barney believed it to be an ordinary object, perhaps a satellite or airplane; but Betty quickly decided that it must be a flying saucer, and she fervently attempted to convince her husband that it was. "Barney! You've got to stop!" she shouted. "Stop the car, Barney, and look at it. It's amazing." (Fuller, 1966: 179) Her near-hysterical excitement proved contagious. Barney stopped the car to get out for a better look, while Betty remained inside ("flopping in the front seat," according to Barney (Fuller, 1966: 146)). He looked at the object through binoculars and fancied that he could see a row of lighted windows with alien faces peering out. The aliens appeared to be busily engaged in pulling levers and quickly turned their backs to him, except for the "leader." Barney was horrified to see that the leader of the aliens appeared to be a "Nazi," as he described it. (Fuller, 1966: 115) Now he, too, was terrified. He got back into the car and drove off. Betty described him as "in a hysterical condition, laughing and repeating that they were going to capture us." (Fuller, 1966: 47) It was at about this point that the supposed "two lost hours" are said to begin.

## What did the Hills Actually See?

The moon was in a region void of any conspicuously bright stars. However, at midnight the planet Saturn was a highly conspicuous first-magnitude object almost directly below the moon, as seen from the White Mountains. But there was a second planet present--Jupiter, two and a half magnitudes (twelve times) more brilliant than Saturn, looking "like a star, a bigger star, up over this one," which parallels Betty Hill's description of the UFO (Fuller, 1966: 171). The Hills reported seeing two starlike objects, one of them the UFO. But two starlike objects were indeed near the moon that evening --Jupiter and Saturn; Mrs. Hill's description of the relative positions and brightness of the objects matches well with the two planets. If a genuine UFO had been present, there would have been three objects near the moon that night: Jupiter; Saturn, and the UFO. Yet they reported seeing only two.

Further evidence that the "UFO" was a distant celestial object is found in Barney's observation that the object stopped when he stopped and started moving again when he did, exactly as the moon or other celestial body appears to "follow" a vehicle (Fuller, 1966: 143). The conclusion is inescapable: no unusual object was present. What Mrs. Hill was calling a

"UFO" must have been the brilliant planet Jupiter. She described it as moving in a saw tooth pattern, a typical description arising from the auto kinetic phenomenon, when the eye tries to follow a bright point source of light against a dark background (Hendry 1979: 26). It is true that Mrs. Hill told John Fuller that, looking in the binoculars, she saw the UFO pass in front of the moon (Fuller 1966: 27), a detail which, if correct, would rule out the object being Jupiter. However, this supposed "fact" was only revealed under hypnosis, and seems to be contradicted by her other hypnosis statement that "all I did was to see it flying through the air and over the front of the car. And you know, I didn't get much of a look at it" (Fuller 1966: 215).

To some it will seem incredible that any seemingly sane person could misperceive a distant (if brilliant) planet as a close-in structured craft, complete with portholes and alien faces peering out. But the examples of numerous other UFO cases prove conclusively that this does indeed happen. Philip J. Klass documents how three educated adults, including the mayor of a large city, observing the reentry of a piece of space debris nearly a hundred miles over their heads, described it as a mysterious craft, with square "portholes," passing less than a thousand feet overhead (Klass, 1974: 9). Allan Hendry's data compiled at the Center for UFO Studies shows that well over half of all so-called close encounters of the first kind can be attributed to either advertising aircraft or stars, with one "third-kind" encounter--an alleged occupant sighting--attributed to each of them. (Hendry, 1979:72-74; 85-86).

Betty's description of the UFO sounds very much like a brilliant point-source of light such as Jupiter: "Even when it was coming in, it still looked like a star. It was a solid-light type of thing ... you couldn't see it too clearly without the binoculars." (Fuller, 1966: 175; 179) Barney was the only one who thought he saw aliens "in real time." After their "abduction," Betty and Barney Hill would go out looking for UFOs, and were to see them many more times. In fact, Mrs. Hill reported that "every night UFOs paced us. Sometimes it was only one, sometimes it was four" (Hill, 1995: 39). This makes it very difficult to avoid the conclusion that what Betty Hill calls "UFOs," other people would call "stars, planets, and airplanes." She acknowledges that some of the UFOs she regularly watches "appear as planes with the same lighting and the sounds of airplane motors" if skeptics are present. "Then I know the night is over, and we leave" (Hill 1995: 147).

# Bad UFOs

A very interesting paper was written in 2007 by **James D. Macdonald**, a former Navy navigator and science fiction writer who lives in Colebrook, New Hampshire, right along the path of the Hills' famous journey, on U.S. **Route 3**. (He is not to be confused with the late atmospheric physicist and UFO proponent Dr. James E. McDonald.) Performing a careful line-by-line analysis of the account of the Hills' Wild Ride in John Fuller's *Interrupted Journey* and also scrutinizing the newer book *Captured* by Kathleen Marden and Stanton T. Friedman (Marden is Betty Hill's niece), Macdonald notes that "the real reason why they were making a forced march is revealed. They didn't have enough money for a motel so they'd decided to pull an all-nighter."

As the Hills passed by tall peaks like Cannon Mountain and went though the Franconia Notch, the Moon and Jupiter would have been too low to see in the west behind the mountains. So surely they must have been looking at something else. But what?

Driving the route late at night, Macdonald noted that "the Cannon Mountain tramway runs 365 days a year, and has been doing so since 1938 when it became the first aerial tramway in North America." A bright light on the lookout tower at the top of the mountain was installed in 1959, and shone all night long (recently removed). Macdonald says,

> Betty would have lost sight of the moon and its accompanying planets as the car went up hill along the side of Mt. Prospect, heading nearly due south. As they crested the rise, the moon and planets would reappear, only now there were two lights to the left of the moon. The light on Cannon Mountain, at that range on a clear night, is as bright or brighter than Jupiter. On a clear night stars appear below the peak of Cannon Mountain to the right and left. Up above we heard the Hills, in a different interview, relate, "it first appeared to be a falling star—only it fell upward." Immediately after cresting the shoulder of the mountain, Route 3 plunges down a 9% grade for the next half mile. The road is pointed directly at Cannon Mountain at this time. Subjectively, at night, I can report of my own direct observation, the light appears to head rapidly straight up...the light on top of Cannon Mountain is visible at various points along this entire route—sometimes high, sometimes low, sometimes to the right of the road, and sometimes to the left... One question that you'd have to answer in order to show this was a flying saucer is, "If

what you saw was a space ship, where was the light on Cannon while all this was going on?"

The Hills reported what appears to be a second Close Encounter with the light atop Cannon Mountain on April 2, 1966: "As we were returning through the Franconia Notch in the general area of the tramway and Cannon Mountain, one [UFO] moved around the mountain about 50 feet from the ground, in front of us. Its lights dimmed out and we could see the row of windows before it became invisible. It just faded out of sight and then just reappeared with different lighting behind us... On the opposite side of the highway was a second one, which also faded out. " (Marden, p. 208-209).

The final spot where the Hills stopped and had a "close encounter" (where Barney says he saw "Nazi" spacemen in his binoculars) was just south of Indian Head. Barney and Betty took UFO researcher Walter Webb to that spot to re-enact the sighting in 1964, and in 2000 Betty Hill took those of us participating in the Indian Head conference there. It's near the location of the now-defunct Mountaineer Motel on Rt. 3 in Lincoln, NH, just north of Exit 33 of I-93 (the interstate was not yet built in 1961). After they drove off from that site, in something of a panic, they never saw the UFO again. Macdonald notes, "2.1 miles south of Indian Head is the last time the Lookout Tower Light is visible from Rt. 3."

As Kathleen Marden describes it, at this point the Hills vaguely remembered "seeing a huge fiery red-orange orb resting upon the ground." Barney relates sitting in the car, motionless, watching an orange light. He says, "I just wasn't driving ahead at this time." (Fuller, 1966: 287) He does not recall how long they sat there, or why he had stopped in the first place. Macdonald writes:

> There's another possible object they may have seen: the Jack O'Lantern Resort in Woodstock which, at the time, had a large billboard with their logo (a stylized jack-o-lantern) down by the road. That would certainly appear to be a "large, luminous moon-shape, which seemed to be touching the road, sitting on end under some pines." This is well out of town; no other features are nearby.

That is a very interesting suggestion. I was not able to find any photo of the Jack O' Lantern resort's giant, unlighted billboard, but I did discover some old postcards with photographs of the giant orange pumpkin that used to sit on the motel roof. Judging from the cars out front, the photo would seem to be from the 1960s. And if the pumpkin were illuminated but the motel lights were out, the "orb" might seem to rest on the ground. In any case, the Hills must have driven past this giant pumpkin on Rt. 3, just minutes after they were frightened out of their wits near Indian Head.

## How Accurate is the Hills' Reconstruction of the Event?

A calibration of the accuracy of the witnesses' descriptions can be made by comparing their description of the weather conditions at the time of the incident (one of their few observations that can actually be checked) with official weather records. This comparison reveals that the Hills' recollection is seriously in error. In their report to the Air Force and in their discussions with John Fuller, the Hills stated that the sky was perfectly clear at the time of the sighting. Yet the official weather station atop nearby Mt. Washington, tallest of the White Mountains, recorded that high, thin cirrus clouds covered more than half the sky at the time of the incident, and this is confirmed by the records of other weather stations throughout New England. Such clouds, while thin and wispy, would be extremely conspicuous in the bright moonlight. Thus we see that one of the few statements in the Hills' account that can he verified turns out to be quite inaccurate. Furthermore, the weather observers on Mt. Washington reported a visibility of 130 miles (!), yet did not report seeing any unusual objects.

## What About the Two "Lost Hours"?

As quoted by Fuller, Barney Hill told Betty while in Colebrook, NH shortly after 10 PM, "It looks like we should be home by 2:30 in the morning—or 3:00 at the latest." Macdonald says, "That's a wildly optimistic estimate. But they'd already decided that they were going to drive home that night. They were on the tail end of a twelve-hundred-mile trip, had run out of money, and were committed to pushing on." In fact, "By the time they reached home they'd been driving for around twenty-one hours. They're lucky that being abducted by space aliens was the worst that happened to them: Others who've tried similar trips have run into trees." Or a deer. Or a moose. In fact, the National Sleep Foundation proclaimed November 6-12, 2011 as its Drowsy Driving Prevention Week®. The AAA's Foundation for Traffic Safety lists as the top "warning sign" of the driver who is too drowsy to drive safely as "The inability to recall the last few miles traveled." The Hills could be the poster children for this traffic safety crusade.

Barney told Dr. Simon that, as the UFO seemed to be moving in closer, "I became slower in my driving...I must have been driving five miles an hour, because I had to put the car in low gear so it would not stall." (Fuller, 1966: 143). The two "lost hours" were said to begin when the Hills drove off after Barney's terrifying face-off with spacemen. However, the Hills' account of that evening's timetable has never been fully consistent. In their report to the Air Force (their earliest account) they gave the time of the reported close encounter as between midnight and 1:00 A.M. In *The Interrupted Journey*, we read that it took place not long after 11:00 P.M. In *The Edge of Reality* there is a transcript of a conversation from a radio show in which Betty Hill says, "We estimated the UFO started to move in close to us right around 3:00 A.M." (Hynek and Vallee, 1975: 9). Which of these times is correct? It is obviously impossible to establish the existence of two lost hours when we have this uncertainty ranging over nearly four hours. We cannot even rely on Barney's oft-cited statement that they left Colebrook at 10:05 PM. As Barney told a church group, "When Betty and I left Colebrook, we never remembered the exact time until we were regressed back under hypnosis. Here, I had looked at a clock on the wall, and the time was five minutes after 10 at night" (Fuller, *Incident at Exeter*, p. 196). Details "recovered" by hypnosis are as likely to be fantasy as fact. If Betty and Barney Hill were unable to give an accurate chronological account of the night's events, how can anyone else hope to do so?

In the Canonical version of the story, a short series of beeping or buzzing sounds was heard, apparently coming from the rear of the car. The first set of beeping sounds caused the Hills to lose consciousness, while the second caused their conscious recollection to return, roughly thirty-five miles down the road. However, Barney Hill told the UFO group APRO that when the beeping sounds began, they kept up for approximately thirty-five miles until they reached Ashland, where the sounds ceased as suddenly as they had commenced (Fitch, 1963). (Apparently Barney didn't know that his memory of this portion of the journey was supposed to be "lost.") Each beep caused the car to vibrate. One possible explanation for the "hypnotic beeps" would be some kind of flaw or corrugation in the pavement, such as often is encountered in construction zones or approaching a toll booth.

Even after the Hills had returned home, it was not evident to them that two hours had been "lost." Not until several weeks afterward, after they had been subjected to intense cross-examinations lasting up to twelve hours by UFO proponents seeking to squeeze from them the last drop of recollection, was it concluded from their inability to account accurately for every minute of the evening that two hours were mysteriously "missing." Macdonald questions the very existence of any "missing time." He asks, "What do they remember south of Indian Head?

a) The Lincoln/Woodstock road marker

b) Downtown North Woodstock

c) (Possibly) the billboard for the Jack O'Lantern Golf Course & Resort in Woodstock

d) Downtown Plymouth

e) Downtown Ashland

f) Entering the superhighway

g) Concord

h) Portsmouth

In short, they remember every single town they passed through. The rest of the trip is past dark lakes, rivers, fields, and woods. I've driven that route more than once, and I don't remember much more than that myself. Not only isn't there any missing time, there aren't any missing memories.

## What does the Hypnosis Testimony Prove?

What about the hypnosis testimony itself? When the late Dr. Benjamin Simon placed Barney and Betty Hill under hypnosis separately, they each told of being "abducted" by alien creatures and then being released with no conscious memories of the incident. This is considered by many to be the strongest evidence supporting the reality of the alleged UFO encounter. However, what a person says while under hypnosis need not necessarily be actual fact. Often it may be a fantasy believed by the individual. Hypnosis is of little value in separating fantasy from fact.

How did the Hills each come to tell essentially the same story? Shortly after the alleged UFO incident, Betty began having a series of dreams about being abducted. She wrote down these dreams and discussed them with her sister and her supervisor. Betty said she never told Barney her dreams, but he was present when she told others. The "abduction" story told under hypnosis was simply a retelling of these dreams. Under hypnosis, Barney admitted to Dr. Simon that Betty had told him "a great many details of the dreams." This prompted Dr. Simon to ask why it was that Barney knew nothing about what had supposedly happened to Betty aboard the UFO, yet Betty seemed to know everything that supposedly had happened to Barney (Fuller, 1966: 237). If Barney was in fact just repeating the dreams his wife had often described, she would of course have far more details of the story than he.

But the most significant fact to be noted about the hypnosis testimony is that *Dr. Simon himself did not believe it*. A careful reading *of The Interrupted Journey* clearly reveals Dr. Simon's skepticism. Dr. Simon indicates what he believes to be the most tenable explanation for the "abduction" story: the dreams of Mrs. Hill had "assumed the quality of a fantasied experience" (Fuller, 1966: 327). When he appeared on the NBC-TV *Today* show on the day that the Hill TV movie *The UFO Incident* would be shown (Oct. 20, 1975), Dr. Simon reaffirmed his opinion: "It was a fantasy ... in other words, it was a dream. The abduction did not happen." Dr. Simon expressed this same opinion to Philip J. Klass, who quotes him at length in two UFO books (Klass, 1974: 252-4; Klass, 1968: 229).

Painstaking experimental research on human memory, taking place in the years since the Hill incident, casts serious doubt upon the entire concept that human memories are formed by something like a photographic process, and that details not at first remembered have nonetheless been accurately stored and can be later brought to consciousness. It is now known that memories are malleable, and that remembering is process of active reconstruction. Later information and ideas are frequently confabulated together with original material, so that an individual may gradually come to "remember" events that never in fact occurred (Loftus and Ketcham, 1996).

**Kottmeyer** observes that Barney Hill's description of his supposed abductors' "**wrap around eyes**" (an extreme rarity in science fiction films), first described and drawn during a hypnosis session on Feb. 22, 1964, comes just twelve days after the first broadcast of an episode of "The Outer Limits" featuring an alien of this quite unique description.

## Is there any Independent Corroboration of the Story?

Another often-cited "proof' of the Hill incident is a supposed "radar confirmation" of the sighting, reportedly proving that an unknown craft did indeed take off at the very time of the supposed UFO abduction. Betty Hill herself has repeatedly made this claim, which has now become widely repeated in UFO circles.

Appearing on the Lou Gordon show in Detroit (WKBD-TV, November 9, 1975), Mrs. Hill alleged, "When the UFO was coming in around midnight, it was picked up on seven different radars, all along the New England coast." Very impressive, if true. In another TV appearance, on the NBC-TV Today show (October 20, 1975), Betty Hill once again claimed that seven different radars had seen the object land about midnight, and she added, "The Air Force in our area released a radar report of it being seen leaving the area at 2:14 A.M.," which appears to confirm beyond any doubt the reality of the alleged abduction. Unfortunately, with just one exception, all the records alleged to support this truly remarkable claim have accidentally been "lost."

Philip J. Klass helped me track down the claims about radar. Klass was also a guest on the Lou Gordon show when Betty Hill repeated her claims about the radar sighting. He asked her if she could provide him with a copy of the documents that are alleged to support them. She readily agreed to do so, but said regretfully that she had only the documents pertaining to one of the radar sightings in her own file--the Air Force radar that reportedly showed the object leaving the area--and that a newspaper reporter had the others. A few weeks later, she informed Klass that the newspaper reporter was unable to provide any documentation supporting any of the seven alleged radar sightings, because records of them had all reportedly been "lost."

The only piece of evidence in existence that in any way supports the supposed radar confirmation of the sighting is a brief paragraph from

Pease Air Force Base in Portsmouth, New Hampshire, that is contained in the Blue Book report on the Hill case: "0614Z observed unidentified A/C come on PAR 4 miles out. A/C made approach and pulled up at 1/2 mile. Shortly afterward observed weak target on downwind, then radar CTC lost. TWR was advised of the A/C when it was on final, then when it made low approach. TWR unable to see any A/C at any time."

Translating the official jargon, the report says: "At 2:14 A.M. E.D.T., an unidentified target was observed on the Precision Approach Radar 4 miles out." (This is a type of radar that sends out its signals in a narrow beam, directly down the runway. It has an extremely narrow field of vision, seeing only those objects that are in a direct line of sight with the runway.) "The object appeared to approach the runway, but left the beam of the radar when it was about one-half mile away. Shortly afterward, another, weaker target was observed, then nothing more was seen. The control tower was twice told about the object, but they were never able to see it."

One highly significant factor is missing from this account--the Airport Surveillance Radar. This is the wide-angle radar that scans the entire region around the airport, keeping track of all aircraft in the area; the Precision Approach Radar merely guides the aircraft onto the runway once the Airport Surveillance Radar has steered it into the general area. As far as the USAF records indicate, the Airport Surveillance Radar saw no unidentified objects at any time. This suggests that the other radar unit was not detecting an actual aircraft when it briefly showed an unknown target. (Sometimes even birds and insects are registered as targets on radars such as the PAR.) The second false target, weaker than the first, confirms the suspicion that the PAR was not detecting any actual craft. Radar "angels," as these false targets are called, give rise to many spurious UFO reports.

Even if we bend over backwards to grant that the Pease radar sighting did indeed represent a genuine UFO near Portsmouth, New Hampshire (along the Atlantic Ocean), there is no reason to connect it with the UFO reportedly seen two hours earlier in the White Mountains, many miles away. Why didn't the observers in the control tower see the UFO if it approached within one-half of a mile? Even more puzzling, why would a UFO enter the runway approach pattern of the Pease Air Force Base, imitating an aircraft about to land?

# What About the 'Alien Star Map'?

Much attention has been focused on this supposed star map, which Betty Hill claims to have seen aboard the UFO and subsequently sketched by post-hypnotic suggestion. The reason for most of this attention has been the work of Ms. Marjorie Fish. Using the most accurate star catalog then available, Ms. Fish strung colored beads on a 3-D frame. She claims to have matched the "stars" drawn by Betty Hill with a group of nearby stars that are all similar to the sun, and which appear to be likely places to find habitable planets.

One of the biggest boosters of the Fish map has been Stanton Friedman, a well-known professional UFO lecturer who bills himself as "The Flying Saucer Physicist" because he worked in physics some 40 years ago, but not since. He writes, "There can no longer be any doubt that the Barney and Betty Hill kidnapping by aliens *did* occur, and we now know exactly where those particular extraterrestrials originated from, thanks to the inspired and intensive research by Marjorie Fish." (Friedman and Slate, 1973) When Friedman and Betty Hill appeared together on the *Tom Snyder Show* on NBC-TV (October 22-23, 1975), Friedman went on to say: "The chances that the Fish map would grab fifteen and come up with the right kind are, well, ... astronomical." "Every one of the stars on the map are the right kind of stars, and all the right kind of stars in the neighborhood are part of the map," he told an amazed Tom Snyder.

Another well-known proponent of the Fish map has been Dr. David Saunders, a psychologist formerly of the University of Colorado and the University of Chicago, who was a dissident member of the Condon Committee. Dr. Saunders, a specialist in statistics, has estimated that the odds against a random pattern of stars matching Betty's sketch as well as the Fish map is "at least 1000-to-1." (Saunders, 1975)

The only problem with statements such as these is that they are incorrect and misleading. All fifteen stars have been identified, according to Friedman, but he neglects to mention that Betty's original sketch contains twenty-six stars, not just fifteen. Why doesn't the Fish map identify the remaining eleven? Three of Betty's background stars (those unconnected by lines) are included in the Fish map, because they fit nicely, but the other eleven are ignored. This hardly constitutes a valid scientific procedure.

As for the claim that all the stars that fit the pattern are exactly the right kind for supporting planets with life, Fish excluded other stars on theoretical grounds, as being unsuitable for supporting life-bearing planets, and hence uninteresting to extraterrestrial travelers.

As astronomers Steven Soter and the late Carl Sagan pointed out (Soter and Sagan, 1975), the only reason that there appears to be any resemblance at all between the Hill sketch and the Fish stars is because of the way the lines have been drawn. View the two patterns simply as dots, without any lines to help the reader visualize the resemblance, and the two patterns look about as different as can be.

Today the Fish Map retains no credibility whatsoever. In her research beginning in 1966, Fish made the wise choice to use the Gliese 2 Catalogue of Nearby Stars, which was then the most accurate available. But that was fifty years ago, and science never stands still. Astronomical researcher Brett Holman recently checked out what the Fish Map would look like if it were built using the most accurate astronomical data available today. His answer is in his article in the British publication *Fortean Times* (#242, November 2008): "Goodbye, Zeta Reticuli" (the supposed home solar system of the UFOnauts). Holman writes, "In the early 1990s the Hipparcos satellite measured the positions of nearly 120,000 stars 10 times more accurately than ever before – including all of those that appear in the Fish interpretation. The results of this work, and much else besides, is available online now, and can be easily queried using websites such as SIMBAD at the Strasbourg Astronomical Observatory."

Fish excluded all variable stars and close binaries to include only supposedly habitable solar systems – but the new data reveals two of her stars as suspected variables, and two more as close binaries. So there go four of her 15 stars. And two more are much further away than earlier believed, removing them completely from the volume of space in question. Six stars of that supposedly exact-matching pattern, are gone, excluded by the very criteria that once included them using the forty-year-old data. Goodbye, Zeta Reticuli.

Also, more than one pattern of stars has been found that appears to match the sketch. At least four other such "identifications" of the Hill sketch had thus far been published. There are simply too many possible ways to interpret Betty Hill's sketch. Random star or planetary positions, when rotated, sorted, and manipulated by choice of position, date, or criterion for inclusion, can be made to match any preestablished pattern, as long as we are willing to expend enough time and effort to obtain a match.

## How Credible are the Principal Witness?

About 1977, Betty Hill began talking about a "UFO landing spot" in southern New Hampshire, where she would go as often as three times a week to watch UFOs. Its exact location was a loosely kept secret, although many reporters and others accompanied her there. Mrs. Hill claimed that Close Encounters are such frequent events at this site that she has made up names for some of the more frequently appearing UFOs: one she calls "the military" because of its allegedly hostile activity, and another is "the working model" (Wysocki, 1977). Betty Hill explains that "the 'new' UFOs aren't friendly as the 'old' ones were. Whereas before they would buzz cars, flying over the roofs and behaving almost playfully, now they sometimes shoot beams, and dart at cars in menacing fashion ... once they even blistered the paint on my car" (Clark, 1978). Mrs. Hill also has asserted that the aliens sometimes get out and do calisthenics before taking off again (Jones, 1978; Hill, 1995: 106).

In a long, three-part article by Dr. Berthold E. Schwartz, a New Jersey psychiatrist formerly with APRO who spent much time interviewing Mrs. Hill, numerous paranormal events allegedly experienced by her are related. She reportedly has encountered a "Pumpkin Head" form that glides beside her car as a UFO hovers above. Afterwards, she is "filled with electricity," setting off airport security devices and resetting electric clocks. She has also performed babysitting chores for a troublesome ghost named Hannah, to give her sister a respite from a long and tedious haunting. After the ghost had reportedly settled in at Betty's home in Portsmouth, "Hannah would walk in the room, cough, and you'd see the rocking chair rock but nobody was in it," Mrs. Hill stated (Schwartz, 1977). Betty Hill has also stated that she believes herself to be receiving telepathic messages from her alien abductors (Fowler, 1976).

John Oswald, a field investigator for CUFOS, accompanied Mrs. Hill on a vigil at her now-famous UFO landing spot, which she described as a "UFO headquarters." He reported that "obviously Mrs. Hill isn't seeing eight UFOs a night. She is seeing things that are not UFOs and calling them

UFOs." On one occasion, he reports, Mrs. Hill was unable to "distinguish between a landed UFO and a streetlight" (Burke, 1977). Nonetheless, Oswald still believed that Betty Hill was abducted by aliens in 1961.

In 2014, skeptics Kitty **Mervine** and Tim **Printy** located Betty Hill's secret "**UFO headquarters**," and paid it a visit. It is at the railroad crossing on Sanborn Road, in East Kingston, New Hampshire (confirmed not only by accounts from those who had visited it with Betty, but also by Betty's photographs of the area). After dark, they saw lights there, too, but using Google Earth determined that they came from the East Kingston Town Hall, located about a mile down the tracks. The building was partly obscured by foliage, which would make the lights seem to twinkle when the wind blows.

From the perspective of the 1960s, a remarkable tale that surfaced via intense interrogation and was later enhanced by hypnosis, containing supposedly recovered memories of extraordinary and remarkable events, might sound exciting and even plausible. However, from the perspective of fifty years later, having seen the rise and fall of hysteria over "repressed memories of rape," "Satanic cults," and "daycare molestation" generated in exactly this same manner, such a tale must be judged implausible in the extreme. Occam's Razor compels us to attribute the complex Hill incident to psychological causes, rather than to "multiply essences" by hypothesizing otherwise-unknown abducting extraterrestrial beings as the cause.

## Betty Hill's Last Hurrah – A Secret UFO Symposium in New Hampshire

One of the most curious events to come out of the Great Internet Stock Bubble was the so-called "Encounters at Indian Head" project, whose very existence was kept unknown to the public for seven years. The symposium was prepared under a shroud of secrecy that was amazingly effective, given the decades-long inability of many top UFOlogists to behave responsibly about anything. Organized by Karl Pflock and the British Fortean author Peter Brookesmith, the event was funded by Joe Firmage, the Silicon Valley then-multimillionaire who seems determined do whatever it takes to bring the public into an even higher state of extraterrestrial awareness.

# Bad UFOs

In September of 2000, I traveled from California to New Hampshire to participate in the secret "stealth" UFO symposium. Firmage was covering all our expenses, and even paid us for the rights to the papers we were writing, which would be published as a book. The purpose of the symposium was, simply, to find out what *really happened* to Betty and Barney Hill. The plan was that nobody would find out about even the existence of the symposium until the book containing its published proceedings appeared 'out of the blue,' presumably creating a sensation. The symposium came off exactly as planned, a tribute to the skills of the late Karl Pflock.

The event was held at the Indian Head Resort, just a stone's throw from the spot where Betty and Barney allegedly saw the UFO cross the road and hover right next to their car. The setting and accommodations were unarguably splendid, the company surprisingly congenial. UFOlogists have a reputation for feuding like Hatfields and McCoys, even with those who they are in general agreement. Probably the high level of the discussion was due to the organizers' careful decision to exclude those UFOlogists who have a reputation for insufferable behavior, whatever their knowledge of the subject. Bravo, Karl. The pre-symposium secrecy ensured that we would not be troubled by the press, the curious, or by certain UFOlogists well known for being pushy and obnoxious. However, the insistence in the non-disclosure agreement for post-event secrecy was more difficult to understand. In January 2001 Pflock announced the "suspension" of the Indian Head project to its participants. The ongoing internet stock collapse undoubtedly cut into Firmage's discretionary spending, with the once high-flying company he founded, U.S. Web (later merged with CKS and March First) bankrupt and liquidated. Still, Firmage paid every cent promised to the participants. With Karl's death in 2006, I feared that the project was defunct, and that the non-disclosure requirement might last indefinitely. Frankly, I now suspect Karl had been dragging his feet and delayed publication of the book for five years because the conference did not turn out as he planned. We spoke on the phone several times about this, and he not only displayed little enthusiasm for completing the book, but implied it would never happen. But after his death Karl's widow, Mary Martinek, quickly completed the editing, and the result is the volume *Encounters at Indian Head* published by Anomalist Books.

*Joe Firmage addresses the particiapnts of the Indian Head conference. From left: Dennis Stacy, Betty Hill, Greg Sandow, Eddie Bullard, Hilary Evans, Peter Brookesmith, Firmage, Karl Pflock.*

The *Grande Dame* of UFOlogy, the late Betty Hill herself, was present to guide us through a re-enactment of the entire "abduction" scenario, assisted by her niece Kathy Marden, who knew the story almost as well as Betty did. I'd met Betty several times before. She regaled us with stories about her literally hundreds of UFO sightings occurring after her initial UFO "abduction." She claims that she organized an entire "Invisible College" of scientists from top laboratories who went out with her to observe and study these UFOs, who gathered reams of documentation and data on the UFOs, then apparently flushed it all down the toilet, as it was their intention to merely *study* the UFOs, and not publish anything about them. Several of the more naïve participants spoke of how listening to Mrs. Hill had made it more difficult for them to accept the reality of her accounts, as if Mrs. Hill's wild stories had not been well-known in UFOlogy for at least twenty-five years. It was the way she told of greeting the extraterrestrials with a jovial "Hi, Guys!" that stuck in the throat of several of the participants. *Not a single participant in the symposium was willing to describe the Betty Hill we heard first-hand as a credible witness;*

nonetheless, a number of them still were inclined to accept her story of alien abduction, including Pflock. The organizers had wisely chosen to send Betty Hill away before we began the actual discussions, as they realized it would be impossible for us to objectively discuss the mental state of a kindly but delusional old lady who was sitting in our midst.

Most of the symposium participants were well known in the UFO and Fortean worlds. Peter Brookesmith of *Fortean Times* magazine, clearly the junior partner as co-organizer, showed himself to be a no-nonsense fellow who also took the partying aspect of the conference very seriously. The good times quaffing with Peter, Karl, and Karl's wife Mary were memorable. Another Brit in attendance was the Fortean Hilary Evans (1929-2011). Two participants were present only virtually: Walter N. Webb, who began a first-hand investigation of the Hill case a month after it occurred in 1961, and Martin Kottmeyer, who wrote amazingly perceptive papers without ever leaving his farm in central Illinois, participated from a distance.

In addition to the conference sessions, we took a Field Trip. First we took the short drive to where the UFO Close Encounter allegedly occurred, just south of the hotel. According to Betty, the Close Encounter site is on the east side of Rt. 3, just north of the present Route 93 freeway interchange (exit 33). She showed us where Barney left the car in the middle of the road with the engine running, while he grabbed binoculars from the trunk to get a good look at the aliens. Betty also guided us to the alleged "capture site", a small, sandy clearing in the woods just off Mill Brook Rd., which goes off NH State Route 175 to the east near Thornton. However, Barney and Betty Hill much earlier had indicated a "capture site" in a different location. [GPS trekkers will want to know that the "capture site" Betty took us to was at 43 deg 54.529' N, 71 deg 39.852' W., elevation 662 ft.] One driver, seeing the small crowd in the woods, stopped to ask if there was a moose on the loose (tourists often travel these back roads seeking Moose Encounters); I replied "no," but didn't have the inclination to explain that we were chasing UFOs. Someone else did, and the driver sped away.

You can learn a lot about a UFO case by visiting the site that you can't learn by reading. Driving from the "Close Encounter" site to the "capture" site, I was surprised to see how many quaint little New England towns lie between them. While driving frantically, allegedly being pursued by the

UFO at close range, the Hills must have driven through the towns of North Woodstock, then Woodstock, West Thornton, and then Thornton. The speed limit in (and around) these towns is 30 MPH. Even granting that these sleepy little towns, which look like they've come out of Norman Rockwell portraits of New England life, would be quiet around midnight, it seems impossible that nobody at all would have noticed a car madly speeding down Rt. 3, screeching around corners, running stop signs and traffic signals, with a low-level UFO in close pursuit. This is related to another great puzzle, to wit: why is it that we never receive reports of UFOs coming in menacingly close, but following *someone else's car?*

We even had an evening screening of relevant science fiction films, including the very episode of Outer Limits that is suggested by Kottmeyer to have inspired Barney Hill's description of the aliens' "wrap-around eyes." There was much discussion of the possible influence of the films on the Hills' account. Firmage sat by himself watching the films, saying nothing. He spent much of his time during the symposium sessions glued to the phone in the hotel lobby, no doubt negotiating major business deals back in Silicon Valley. His participation was slight. I did have a chance to speak with him for a few minutes during the first evening session. He confidently expounded about how one dissident physicist or another had come up with a theory showing that it is possible to do the things that UFOs allegedly do: travel faster than light, defy gravity, etc. For him, this settled the matter: such things were possible, and we should drop our present-day prejudices. He seemed not to appreciate the objection that the great majority of physicists were unconvinced by unsupported speculative theories, or else he seemed not to care. Firmage is an impressive, dynamic speaker, but not such a good listener.

Ultimately, no agreement was reached concerning whether the Hills' story was real or imagined. Each participant (except for Greg Sandow) expounds his viewpoint at length in a chapter in the book. Eddie Bullard, Greg Sandow, Walter Webb, and Karl Pflock argued that the Hills' abduction account should probably be taken literally. I argued strongly for the opposite, as did Peter Brookesmith. Martin Kottmeyer and Hilary Evans agreed that the explanation was more likely to be psychological than physical. Dennis Stacy, a former editor of the MUFON Journal and the publisher of the symposium volume, limited himself to carefully chronicling and recounting the incident. However, in private conversation he confessed to difficulties with accepting the Hills' account. Sociologist

Marcello Truzzi pronounced it impossible to come to any conclusion whatsoever.

It was clear that the participants who had not previously met Betty Hill were dismayed and/or disappointed after hearing her ramble on glibly about things that could not possibly be true. However, there were rationalizations aplenty as to why we should believe her claims made in 1961, but not afterwards. I also felt that co-organizer Karl Pflock, and sponsor Joe Firmage, had expected some sort of pro-Hills consensus to emerge from the discussions when all of the "facts" supporting it were marshaled – and were rather disappointed when it did not. One of the "evidences" in favor of the alleged abduction has long been Betty's statement that her husband Barney, after having his genitals examined by aliens, developed a ring of warts around his groin. The pro-abductionists seemed genuinely startled to be told (after Betty had safely departed) that this symptom is evidence, not of alien activity, but of a common venereal disease.

## The Influence of the Hill Case Upon UFOlogy

Prior to about 1966, when the Hill abduction story became well known to the public, accounts of "UFO abductions" were virtually unknown. The Hill case, with its seemingly high degree of credibility, broke new ground in terms of its acceptability to the UFO movement. It was the first claim of face-to-face contact with UFO aliens that was embraced by the mainstream of the UFO movement. NICAP - the largest and most influential UFO group then in existence - had previously rejected all cases in which direct contact with extraterrestrials had been claimed. The Hill case turned out in retrospect to have been a watershed event, opening the floodgates for a torrent of subsequent "UFO abduction" claims, contributing significantly to the climate of ever-increasing credulity that has characterized the UFO movement.

A flood of "abduction" stories followed in the wake of the NBC-TV telecast of the movie dramatization of the Hill case *The UFO Incident* on October 20, 1975. A minor "UFO flap" in certain areas of the country occurred as well.

A significant portion of the book *Abducted* by APRO's James and Coral Lorenzen consists of spin-off cases generated in the wake of the movie.

Some of the cases derived from alleged sightings from a few months earlier, whose "missing hours" and "abduction" aspects were only recognized after the supposed abductee watched *The UFO Incident* on October 20. The Lorenzens write "the summer and fall months of 1975 appear to have been a busy time for whoever and whatever was generating the reports of abduction" (Lorenzen and Lorenzen, 1977). That the Hill TV movie and its related publicity may have been the reason seems not to have occurred to them. Among the best-known spin-off cases were Mrs. Sandra Larson of Fargo, North Dakota; two young men in Norway, Maine; the celebrated "Fire in the Sky" Travis Walton abduction; and Judy Kendall, a legal secretary living in northern California. In addition to the spate of "abduction" cases, the Hill movie seems to have triggered a respectable, if not overwhelming, flap of UFO sightings. A Defense Department memo, dated November 11, 1975, declassified under a Freedom of Information Act request, reads "Since 28 OCT 75 numerous reports of suspicious objects have been received at the NORAD COC (Combat Operations Center)..." (NY Times Magazine, Oct. 14, 1979). The reports originated from military personnel at bases in the northeastern United States and adjacent regions of Canada. Note that the "flap" began exactly eight days after the Hill movie was shown. In the period of November 4-9, a UFO "flap" reportedly associated with cattle mutilations broke out in Wisconsin, and a police sighting was reported in Ohio.

The late Richard Feynman, the Nobel prize wining physicist, had this to say about UFO cases: "almost everybody who observes flying saucers sees something different, unless they were previously informed of what they were supposed to see" (Feynmann, 1998). The historical significance of the Hill case is that, because it was the first account of its kind, and because of the massive publicity that followed in its wake, it informed those who came later what they were supposed to see.

## Ripple Effects from the Hill Case in the Culture at Large

Starting in the 1980s, dramatic events discovered via supposedly "recovered memories" began to receive widespread attention. When these cases began being reported, the only precedent in the culture at large for "recovered memories" was in UFO abduction reports. A small number of women began "recovering memories" of supposedly being raped during childhood, usually by a family member. Strongly embraced by the feminist movement, the original few grew into a small torrent.

Prodded by the best-selling *The Courage to Heal* (Bass and Davis, 1988), millions of women worldwide imagined themselves to be "recovering memories" of supposed sexual abuse. Soon "recovered memory therapy" was all the rage in therapy. Thousands of families were devastated by mostly-groundless accusations, and few gave such claims the critical examination they required.

Around the same time, lurid accusations began circulating about organized child sexual abuse that was supposedly occurring nationwide in daycare centers. Often supposed "satanic cults" were claimed to be involved, with lurid claims of victimization ranging from the implausible to the completely impossible. A number of innocent persons spent years in prison, usually because of bizarre testimony coerced from small children. The fad grew quickly until the mid-1990s, when several persons whose lives had been ruined by false accusations of rape or "satanic rituals" fabricated during therapy had successfully sued for damages the "therapist" who instigated the supposed "recovered memories". With "therapists" suddenly being held accountable for the harm resulting from the delusions they had implanted, the fad for "recovered memory therapy" faded as quickly as it had begun.

In Elaine Showalter's study of modern hysterical epidemics, in which "alien abductions" are featured as a prime example, the delusion is considered in a broader societal context. Noting that the great majority of supposed UFO abductees are female, as are 90% of those who claim to have recovered memories of childhood sexual abuse, and 90% of those who are diagnosed with Multiple Personality Disorder, she views the hysteria as serving the function of a kind of "women's protest against patriarchy." (Showalter, 1997: 10). Showalter argues that if we want to understand the cause of these hysterias, we cannot do so by ignoring the social context in which they occur, and the "empowering" role they play.

## November, 1975: Travis Walton – the "Fire in the Sky"

After the Hills, the best-known UFO abduction story is that of Travis Walton. He claims that on Nov. 5, 1975, as he and six others were returning from a day's work cutting logs in Arizona's Sitgreaves National Forest, he was zapped by a beam of light from a UFO, and taken aboard the craft for five days. His story won $5,000 as the Best UFO Case for 1975 from the *National Enquirer*, has been the subject of several books, as well as the Hollywood movie *Fire in the Sky* (1993), and the documentary film *Travis: The True Story of Travis Walton* (2015).

Philip J. Klass wrote many letters and made many phone calls to people involved in the Walton story, including Travis' family, the local Sheriff's office, polygraph examiners, etc. He found strong reasons to brand the entire story a hoax. His conclusions were written up in great detail in chapters 18-23 of his book *UFOs The Public Deceived* (Prometheus, 1983). Still, the skeptic's case against the Walton abduction claim is not being widely read. For several years I have had a Travis **Walton** page on my *Debunker.com* website, but not as much information as Klass provides.

In his 1976 "**White Paper**" about the **Walton** abduction claim (on my *Debunker.com* website), Klass summarizes the case as follows:

> On the evening of November 5, 1975, at approximately 6:15 p.m. MST, a crew of seven young wood-cutters, headed by Michael Rogers, was returning home. Rogers (age 28) was under contract to the U.S. Forest Service to thin out 1,277 acres of National Forest land near Turkey Springs. According to the story later told by Rogers and members of his crew (ages 17-25), they saw a UFO hovering nearby. They claim that Travis Walton jumped out of the moving car and walked/ran under the UFO, that he was "zapped" by an intense glowing beam from the UFO and that the rest of the crew panicked and drove off, leaving their friend behind. A short time later, they said, they returned to the spot to seek Travis but that he had disappeared -- seemingly having been carried off by the UFO. However, it was not until more than two hours later that Rogers and his crew reported the incident to Under-Sheriff L.C. Ellison in nearby Heber, Arizona.

The first UFO investigators on the scene were from a pro-UFO organization called Ground Saucer Watch (GSW), arriving to interview Travis' family and friends while he was still "missing." The GSW investigators immediately became suspicious of the case. Among their findings:

> · The entire Walton family has had a continual UFO history. The Walton boys have reported observing 10 to 15 separate UFO sightings (very high).

· When Duane was questioned about his brother's disappearance, he stated that "Travis will be found, that UFO's are friendly." GSW countered, "How do you know Travis will be found?" Duane said "I have a feeling, a strong feeling." GSW asked "If the UFO 'captors' are going to return Travis, will you have a camera to record this great occurrence?" Duane, "No, if I have a camera 'they' will not return."

· The Walton's mother showed no outward emotion over the 'loss' of Travis. She said that UFO's will not harm her son, he will be returned and that UFO's have been seen by her family many times.

· The Walton's refused any outside scientific help or anyone who logically doubted the abduction portion of the story.

Klass continues:

While Travis was missing, Rogers and the other five young men took a polygraph test (on Nov. 10), administered by C.E. Gilson of the Arizona Dept. of Public Safety. Five of the young men "passed" the test but the results for one (Allen M. Dalis) were "inconclusive" according to Gilson. These test results have been widely interpreted as endorsing the authenticity of the alleged UFO abduction.

Shortly after midnight on Nov. 11, Travis telephoned his sister, Mrs. Grant Neff of Taylor, near Snowflake, from a phone booth in Heber, about 30 miles away. Mr. Neff and Travis's older brother Duane, who had come to Snowflake from his home in Phoenix shortly after, the alleged incident, both drove to Heber to pick up Travis. They reported finding him crumpled on the floor of the phone booth, in a very "confused" mental state. A short time after returning Travis to his mother's home in Snowflake, Duane decided to drive Travis to Phoenix, reportedly to obtain medical assistance. Later that same day, he was examined by two physicians at the request of [UFO group] APRO.

On Feb. 7, 1976, almost three months after Travis's return, he and Duane took polygraph tests administered by George J, Pfeifer, then employed by Tom Ezell & Associates of Phoenix. According to published reports, both men passed the exam which involved many questions dealing with Travis's claim of having been abducted by a

UFO. The widely publicized results of these tests seem to confirm that such an incident actually occurred.

WHAT THE PUBLIC AND APRO MEMBERS HAVE NOT BEEN TOLD AS OF THIS DATE (6/20/76) IS THAT TRAVIS WALTON TOOK A LIE-DETECTOR TEST ON NOV 15, 1975. HE FLUNKED IT!

...Immediately after the test, [examiner John J.] McCarthy reported his findings of "gross deception" to Paul Jenkins of the "National Enquirer" and to Dr. James A. Harder, APRO's director of research and Harder then relayed the results by telephone to APRO's Lorenzen, according to McCarthy. Duane Walton, upon hearing McCarthy's conclusions, became furious with the polygraph examiner, McCarthy told me.

So, the *National Enquirer*, and the then-prominent UFO organization APRO, were informed by their polygraph examiner, trained by the U.S. Army and with 20 years' experience, of Walton's "gross deception," but chose to "deep six" the damaging information. One member of the *National Enquirer* team that worked on Travis Walton case, **Jeff Wells**, later wrote a very revealing "insider's view" of the shabby and deceptive efforts taken by Harder, APRO, and others to successfully protect the damaged brand of the "Walton abduction" and re-package that valuable property for an unsuspecting public. Wells' account, titled **Profitable Nightmare of a Very Unreal Kind,** was first published in *The Age*, Melbourne, Australia, on 6 January 1979. It was later reprinted in *The Skeptical Inquirer*, and is on my **debunker.com** website.

## A hoax is a hoax of course, of course, go right to the source and ask the horse

The motive for the hoax, according to Klass, was that Mike Rogers was seriously behind on his contract with the Forest Service, and was desperately seeking an excuse for getting out of it.

As the Forest Service's Nov. 10 deadline approached, it was clear to Rogers that he could not meet even his already-extended contract date. In the Nov. 8 taped interview with Sylvanus [one of the GSW investigators first on the scene], Rogers mentioned that "this contract we have is seriously behind schedule, in fact, Monday the time is up. We haven't done any work on it since Wednesday because of this thing, and therefore it won't be done. I hope they [Forest Service] take that into account, this [UFO] problem." Without the 'fortuitous' UFO incident, the best that loggers could expect would be another deadline extension, but he would be assessed still another financial penalty in  the contract's price.

*Note that this concern over not finishing the contract was expressed even while Travis was still "missing," and possibly being fricasseed by alien beings.* Yet his good friend Mike Rogers is worried about his Forest Service contract, not about the missing Travis. If Rogers can persuade the Forest Service that his men are too frightened to go back into the woods to cut trees, he might hope to avoid the financial penalty.

Assuming that a hoax was involved, exactly who was hoaxing whom? Klass always assumed that all six woodcutters in the truck were part of the hoax, a supposition that has often been criticized as implausible, especially since several members of the crew had only recently met and did not know each other well. As explained by John Harney in "If You Go Down to the Woods Today" (*Magonia* 74, April 2001):

> The men must presumably have been intelligent enough to realise that when they reported to the police that Walton had disappeared, they would be subjected to close questioning, not only by the police but by journalists and ufologists. It is difficult to imagine how such an unlikely concocted story could stand up to such pressure. Their first test came when they drove into Heber and phoned the police. When Deputy Sheriff Chuck Ellison met them there and was later joined by Sheriff Martin Gillespie and Undersheriff Ken Coplan. The men were said to be in a highly emotional state. At least, so far as I know, no one has denied this... This means, of course, that if they were all in on the hoax they were displaying considerable acting talent. Their performance would surely have involved much rehearsing as well as histrionic ability.

But  Karl Pflock had a different interpretation of the Walton hoax, and one that I believe is correct. Harney, who interviewed Pflock, continues,

> If Rogers and Walton devised the hoax, then they must have rigged up something in the forest which they could use to fool the other men into believing that it was a flying saucer. One advantage of this

142

interpretation of the story is that, when the men were being interviewed by the police officers, it was only Mike Rogers who was putting on an act and the others were telling the truth, and their emotions were perfectly genuine...

if we suppose that Walton and Rogers managed to deceive the other five men then there is no problem here. The weak point comes at an earlier stage, when a strange light is seen as the men are driving out of the forest. Assuming that something has been rigged up in the trees and suitably illuminated, then the difficult bit is in Rogers and Walton managing to fool the other five men into thinking it is a flying saucer. Also, Rogers has to ensure that the other men stay in the truck [only Rogers and Walton are seated next to a door] and that he does not hang around long enough for any of them to make out what the object really is.

The object is seen ahead of the truck and to the right and it just so happens that Walton is sitting on the front seat on the right. The object also has to produce a bright beam or flash of light at the right moment, so that Walton can be "zapped" and fall over backwards, apparently unconscious. This is supposed to cause the men to panic and they obligingly do so. Rogers then drives off so that they don't get a chance to see what happens next.

In 1978 Klass interviewed woodcutter Steve Pierce by telephone, and asked him about the supposed UFO:

Klass: What did you see?

Pierce: Uh, well, I thought it was something a deer hunter, you know, rigged up. You know, 'cause it was deer season, you know, so he could see. You know? And, uh, and, but I couldn't see the bottom or a top or sides, all's I could see was the front of it, you know. You couldn't tell if it had a bottom to it or, you know, or a back to it or anything...

About which Pflock commented, "Hmmmmm... A 'Plan Nine from Outer Space' saucer, perhaps?" Seriously, the flying saucer looked like a deer hunter's blind?

Klass also writes that Pierce told him that, on the day of the "UFO abduction," Walton did no work at all, claiming to be ill. During the afternoon, Mike Rogers disappeared from the work site for about two hours (perhaps to prepare the UFO light show). They usually left the work site about 4:00, according to Pierce, but on that day they remained until about 6:00, thus they were uncharacteristically driving home in the dark. If they drove home in daylight, the UFO light show would not be visible, and the real identity of the contraption could more easily be determined.

## UFO Bribe or Bluff?

Speaking at the International UFO Congress in Arizona in February, 2012, Travis Walton brought out, for the first time, two of the men who were present when he was "abducted," Steve Pierce and John Goulette. Travis made a big thing about an accusation that Philip J. Klass had supposedly offered Pierce a bribe of $10,000 to say that the 'abduction' was a hoax.

That accusation originated not with Pierce himself, but with Travis' pal Mike Rogers back in 1978. In Bill Barry's 1978 book about Travis Walton, *Ultimate Encounter*, it says

> According to Mike Rogers, "Steve told me and Travis that he had been offered ten thousand dollars just to sign a denial. He said he was thinking about taking it." (p. 160)

However, note that the accusation does not come from Pierce, but instead from Mike Rogers. Barry does not directly accuse Klass of offering a bribe, but hints it is so. In that book Rogers told Pierce, "Then you'll spend the money alone, and you'll be bruised." Klass writes, "The latter suggests that Rogers was threatening Pierce with physical harm if he recanted." If you read reporter Jeff Wells' *A Profitable Nightmare of a Very Unreal Kind* (on **debunker.com**) you would see that such a threat of violence should be taken seriously. Klass continued (Klass 1983, p. 221), "had Barry checked with me, I would have assured him that I never made such an offer to [Deputy] Click or to anyone seeking to 'buy off' a member of the Rogers crew.

When Klass read the Barry book, he phoned Pierce - after the accusation had already been made - who told him that he thought the abduction was a hoax, but he could not prove it. Note that Klass did not travel to Texas or anywhere else to meet with Pierce, or take him out and entertain him -

those are lies made up later by the Walton crowd. Klass recorded that phone interview, as he did every significant interview, and later sent a copy of the transcript of it to Karl Pflock upon request.

But by 2012, Pierce's story had been brought into line (otherwise Walton would never have included him in the presentation): 'absolutely yes, Klass tried to bribe me. He flew out to Texas to wine and dine me and try to persuade me. He kept following me, I had to move to like three different states, to get away from him.' Of course, there is no proof that this 'new version' of Pierce's story is correct, no photos of Klass and Pierce together, no documents of any kind to back up this implausible tale. Was there any documentary evidence to either support or refute this?

Klass willed his papers to the American Philosophical Society in Philadelphia, PA, where the collection now resides. I contacted them, requesting any files in Klass' Travis Walton case files concerning Klass and Steve Pierce.

The result was a PDF file that completely supports in every way what Klass wrote about his brief interactions with Pierce in his 1983 book *UFOs The Public Deceived* (see **debunker.com**). It is a transcript of the 1978 phone conversations between Klass and Pierce. At no time does Pierce suggest that the "bribe" story is true. In fact, he told Klass

> If I ever could prove it was a hoax I'd damn sure do it.

Interestingly, Pierce says some rather unflattering things about Travis Walton. At one point, Pierce tells Klass,

> Travis is the most ignorant, stupid person I've ever met in my life. He ain't got enough sense, you know the book he wrote, he couldn't have wrote that book by himself. He ain't got enough common sense to write that book. [It was widely rumored that John G. Fuller, author of *The Interrupted Journey* and other books, was Walton's ghost writer.]

# 'Bedroom Abductions' Replace 'Deserted Road Abductions'

The face of UFO abductology changed dramatically with the 1981 publication of *Missing Time* by Budd Hopkins (1931-2011). The earliest reported UFO abductions in the U.S. - Betty and Barney Hill in 1961, and a trickle of others including Travis Walton in 1975 - typically involved going outside to some lonely and deserted spot at night, where one allegedly encountered aliens, and was kidnapped. It was Hopkins who severed that connection completely in the early 1980s. No longer was it necessary to be outside in some scary place at night for a UFO abduction to occur: in the new Hopkins-style abductions, the aliens would come right into your bedroom and snatch you up, often passing through solid walls in the process. According to Hopkins, if your kid calls out "Mommy, there's a monster under my bed," the kid just might be right. We are all familiar with the idea of people supposedly being abducted by aliens from UFOs. It is now a commonplace in popular culture. In 1992, CBS-TV produced a prime time mini-series based on UFO abductionist Budd Hopkins' writings, titled *Intruders*.

A typical "UFO abduction" scenario from the 1960s through the early 1980s runs like this: a sighting of an unidentified object; wondering about a possible abduction; regressive hypnosis leading to "discovery" of a previously forgotten memories of physical examination aboard a UFO by alien beings; finally, subsequent experience of paranormal events. The debt to the Hill case is obvious. Several major "abduction" cases were reported before the Hill movie was telecast, but had already been widely disseminated in print. Patty Price first believed she may have been abducted after reading a *Saga* magazine article on UFOs. Sgt. Charles I. Moody first speculated he may have been abducted after happening upon a copy of *Official UFO* magazine (Lorenzen and Lorenzen, 1977).

The "UFO abduction" scenario changed significantly after the arrival of Budd Hopkins. The alien abductors became more sinister, their intentions frankly more sexual. Hopkins wrote of a new "abduction scenario that is, if anything, even more disturbing to contemplate. Many people have simply been taken from their homes while they were either asleep, or engaged in some quotidian activity, like watching television or reading" (Hopkins, 1981: 79). Hopkins believes that the space aliens are conducting genetic experiments on humans. When the view of UFO abductions promoted by

# UFO Abductions

Hopkins (along with his colleagues David Jacobs and John Mack) became the dominant paradigm in UFOlogy, the Hill case began to fade in significance. Henceforth, major UFO abduction cases almost invariably tended to fit the Hopkins pattern rather than that of the Hills (which in itself strongly suggests the entire "abduction" scenario to be of psychological origin).

In her privately-published book *A Common Sense Approach to UFOs* (the irony in the title is delicious!), Betty Hill strongly rejects the new Hopkins paradigm for UFO abductions, arguing that such cases are psychological in origin. Her arguments against relying upon hypnosis-enhanced "memories" are surely valid, seemingly forgetting that her own case depends upon them, too. Mrs. Hill writes, "Real abductions have the same characteristics. The event happens in a rural, isolated area usually between dusk and 4 AM" (Hill, 1995: 81).

The Hopkins/Jacob/Mack theme of repeated abductions, typically beginning in childhood, establishes a personal relationship between abductee and abductor. No longer is abduction simply the result of being in the wrong place at the wrong time, as was supposedly the case for Hill-style abductees. Instead, the abductee (overwhelmingly female) became a special kind of person with a mystical, cosmic lifelong bond connecting her to unknown cosmic forces and beings. Most current abduction stories contain claims that rudely violate common sense even beyond the dubious idea of alien visitations. Often the creatures and the abductee are said to levitate or fly, to pass through solid objects, or to simply teleport themselves from one location to another. Hopkins has actually suggested, in all seriousness, that the aliens have the ability to make themselves and their victims invisible to better preserve the stealth of their operations. Given such claims, plus the frequently sexual nature of the abduction experience, the correlation with dream states and sleep disorders is obvious.

By the early 1990s, abduction mania had become a significant social phenomenon. Abductology was riding high, led by its Troika of Hopkins/Jacob/Mack. Speaking to the *Bay Area UFO Expo* in 2000, Budd Hopkins talked about how abduction experiences affect young children. In his workshop, which costs extra to attend, he promised to discuss the signs by which childhood abductions can be realized. Hopkins talked of "Stages of an abductee's Psycho-social development." In some cases, a newborn baby has been abducted from the hospital before its mother took it home. If babies show fear of its mother wearing dark glasses, or fear of being put into the crib at night, it suggests alien abductions. A styrofoam wig-stand was painted with alien eyes by a young girl. It fascinated her, and frightened her younger siblings. Some alien implants, once noted, will spontaneously disappear. This happens all the time. Young abductees in school tend to be loners, and feel inadequate. Barney and Betty Hill had probably been gotten by aliens many times before their famous abduction, according to Hopkins.

Speaking at that same conference, David Jacobs painted a scary scenario of clandestine activity by one or more alien races, whose agenda is "secret." Fortunately, he has been able to penetrate many of their secrets owing to the recollections of abductees, despite the aliens' best efforts to keep their activities secret. Apparently their advanced knowledge of human neurophysiology enables them to largely prevent abductees from remembering all aspects of their ordeals.

Jacobs is obsessed by the "tables" in the saucers, where the examination of abductees supposedly takes place (recall the abducted Wilma strapped to a table in the Buck Rogers comic book abduction by Martians). These "tables" are not merely placed on the floor, but instead are built into the structure of the object. They are obviously built to accommodate humans, not the little guys who pilot the saucers. And there are lots of tables. Indeed, there are rooms with nothing but tables. These tables were obviously manufactured by the ETs for a purpose. And there are a great number of UFOs. Virtually everything that is inside the UFOs seems to be constructed for the purpose of the abduction phenomenon (i.e., there are no "rec rooms" for the ETs). Jacobs is suggesting that the ETs are pursuing their mass-abduction agenda with single-minded ferocity.

The insectoid creatures, often described as Praying Mantises, are clearly the order-givers. They are obviously in charge. Some people describe Reptilians, although it is possible that they may be just other descriptions of the Praying Mantis. [Insects and Reptiles are, after all, easily confused.] The hypothesis that the ETs are conducting "experiments" is no longer tenable, said Jacobs. A vast, clandestine program of alien activity is underway.

Hopkins and his colleagues were so confident about the "scientific" status of their findings that in 1992 they arranged an Abduction Study Conference at MIT, hosted by physicist David Pritchard, in which I participated representing CSICOP. However, they went to extraordinary lengths using "non-disclosure forms" to control how the conference was reported. While the participants were heavily slanted toward the pro-abduction view, there was a significant presence of skeptical professionals, and instead of solidifying the abductionists' claims, the conference highlighted their glaring weaknesses

The conference did not unfold as smoothly as its organizers planned. Many academics, even those inclined toward UFO or paranormal belief, objected mightily to the loose "methods" of the Troika. In one of Budd Hopkins' talks, he described a survey he did of children showing them pictures of unusual things to see which ones they were familiar with, to tell if they might have been abducted. He was met by an avalanche of objections: you didn't normalize, you didn't validate, etc. In other words, his survey was worthless. Chastened, Hopkins said something like "I'm sorry, I'm just an artist and I don't understand all that technical stuff. I thank you guys. That's why we invited you here, to help us." Not long afterward, Mack was speaking and described another test or survey he was doing with his subjects. He ran into similar objections. I was waiting for Mack to say, "I'm sorry, I'm just a Professor of Psychiatry at Harvard University, and I don't understand all that technical stuff." But he did not.

Hopkins and his colleagues used the conference to first reveal details of a spectacular alleged multiple-witness abduction case that occurred late one night in Manhattan. This case became the subject of Hopkins's 1996 book *Witnessed: The True Story of the Brooklyn Bridge Abduction*. However, independent pro-UFO researchers were unable to confirm the ever-shifting claims about multiple witnesses to the alleged abduction, and the abductee's ever-spiraling and ever-changing tales of encounters and intrigue became increasingly difficult to believe. The case that Hopkins and his colleagues had "bet the house" on, expecting it to finally establish their claims, ended up as a humiliation.

## Whitley Strieber and 'The Visitors'

In 1987 Whitley Strieber, already well-known as a writer of fantasy and science fiction for *Wolfen, The Hunger*, and other exciting tales, published his best-selling supposedly non-fiction book *Communion: A True Story (1987),* about his ongoing encounters with ET-like beings he calls "the visitors." It was on the *New York Times* bestseller list for many weeks, followed up by *Transformation* in 1988, and then by numerous other paranormal-themed books, some admitted to be novels, others not. Imagine if Steven King had written a masterful horror novel, but claimed that all of the ghostly goings-on really did happen to him, although he could offer no proof of any of it. That's where we stand with Strieber.

Whitley Strieber created a new element in UFO lore with *Communion*, which was made into a Hollywood movie in 1989. Strieber claims that the "visitors" gave him an anal probe:

> two of the stocky ones drew my legs apart. The next thing I knew I was being shown an enormous and extremely ugly object, grey and scaly, with a sort of network of wires on the end. It was at least a foot long, narrow, and triangular in structure. They inserted the thing into my rectum. It seemed to swarm into me as if it had a life of its own. Apparently, its purpose was to take samples, possibly of fecal matter, but at the same time I had the impression I was being raped and for the first time I felt anger.

*An alien device protrudes from Cartman's rectum and he farts fire after an alien abduction on "South Park".*

He also noted that the female humanoid was very interested in his penis.

In 1997 the very first episode of the first season of the enormously popular satirical animated TV series *South Park* was titled "Cartman Gets an Anal Probe." In it, the charismatic young narcissist Eric Cartman is abducted by a UFO, and is implanted with an anal probe consisting of a huge satellite-type communications system jammed into his rectum, that sometimes protrudes out a meter or more out from his fat ass to "phone home" to the UFO. A 1995 skit on *Saturday Night Live* also dealt with alien anal probes.

# UFO Abductions

I had my first Close Encounter with the Great Confabulator on September 21, 1988. Viewers of the popular daytime television show *People Are Talking* on KPIX, Channel 5, in San Francisco saw an amazing thing. Whitley Strieber indignantly refused to let the hosts of the show do any promotion of his latest book! No doubt the viewers of that show were scratching their heads about such inexplicable behavior on the part of a guest doing a book promotion tour. As the other guest on that show, the one who was all but ignored by the hosts, let me explain why that strange scene happened.

Forty-five minutes before air time I arrived at the studio and was escorted to the Green Room, where guests are groomed and prepared. There I came upon Whitley Strieber in the midst of a world-class temper tantrum. He was indignantly refusing to go on! He apparently expected to be the only guest, and to have an entire hour to expound his fantasies about the humanoid "visitors" who are said to be lavishing their unwanted attention on him, unchallenged and unquestioned. I later found out that while he had left instructions with those arranging his book tour that under no circumstances would he appear on any show with Philip J. Klass, he had not ruled out - at least to them - appearing with some other skeptic. But here he was, hurling verbal assaults. "I don't know who this man is, and I don't know what he will say!" Apparently he expects others' opinions to be cleared with him in advance.

Whitley continued his tirade. Pointing to me, he shouted "that man is going to go on and challenge my mental health. He's going to call me crazy! He's with that CSICOP, they're just as nuts as those new-age people. They have a religion of disbelief." In his short tirade against the skeptics, who he says are in the habit of calling anyone who disagrees with them crazy, Strieber called us "nuts" or "crazy" three times. I pointed out the irony of this, but it was clear from the reaction of all involved that the best thing I could say at this point was nothing. I kept silent for a while, enabling him to resume his tirade. He had received long letters from Philip J. Klass of CSICOP, he said, that were "crazy," and made no sense at all. He also charged that the hosts of the show were bound to misrepresent his experiences by saying that they are alien visitors, while he has never claimed to know whether or not "the visitors" are extraterrestrial. Those people who claim alien encounters are just as crazy as CSICOP, he charged. "I don't need your show," he continued, "your stupid show! My book (*Transformation*) is number four on the Best-Seller list. I don't need to do these shows! I'm getting so fed up with going on shows and having everyone laugh at me!" The producer emphasized that a live show would be starting very soon, on which he had agreed to appear, and that he must meet his commitments. But Whitley still refused to go out and appear or debate with me. "Let him go on first. I'll just do the final segment. And DON'T mention my book! I don't want you to mention my book at all if he is going to be criticizing it!" Then Strieber must have realized that he couldn't win this battle. He gradually decreased his level of objection, the bluster slowly fading as it became clear that he was not going to be able to keep me off the show. "All right," said Whitley, "I will go on - but I WON'T LIKE IT!" The magnitude of that threat stunned all who were present. "And I'll never come back!"

They kept me off camera much as possible. During the commercial break before one of the final segments, the producer dashed out onto the set to ask Strieber if he wanted his book to be "promo-ed". "NO!" he flatly replied. I said that I would like to have *my* book, *The UFO Verdict*, "promo-ed". Whitley said, still annoyed, "Yes, go promo *his* book!" This was done, briefly. In the

*Whitley and the late Anne Strieber in 2012*
*(photo by author)*

final fifteen seconds of the show, the host asked Strieber from across the room if he wanted to mention his book. "NO!", Whitley snarled, then paused, and sheepishly muttered, "it's *Transformation*." Within seconds of going off the air, Strieber had left the studio. The 'Prima Donna' was still furious.

## He Sees Dead People

When I first saw Strieber with his wife and collaborator Anne seated on a panel at the 2012 International UFO Congress, I remarked to myself how their demeanor reminded me of the couple in the famous painting

American Gothic. I had not seen Anne before. (Anne Strieber died in 2015.)

I got out of my seat to get photos of the speakers, without a flash so as not to disturb the video being made. I was looking at the photos as I returned to my seat. The fellow next to me asked "were there any orbs above them?" "No, not this time," I replied. Strieber returned to the theme of contact with the dead that he first began to promote in his book *The Key*. Dead people, he says, dress in a brown monk's cowl, "Jesuit clothing." Some of the visitors think it is possible for mankind to Evolve, while others apparently are not so optimistic. That is why the visitors are so cautious and stealthy. Afterward, mankind will be changed completely, and come face-to-face with The Dead. Strieber still says he does not know exactly who "the visitors" are, but he knows that The Dead play a large role.

At a MUFON mini-symposium in Costa Mesa, California in 2012, Strieber was on hand to explain that he has an implant in his left ear. After a particularly spooky encounter one night with The Visitors, Whitley felt a "nasal intrusion," and later realized that he had been given an implant in his ear. As he writes in his book *Solving the Communion Enigma*, "later that day, I noticed a sore lump at the top of my left ear... from time to time the ear would get red and hot." He went to a doctor who diagnosed a "probably benign cyst" that was "fixed in the pinna of my ear, close to the top," and set out to have it removed. However, the implant would not allow itself to be taken alive:

> He started to draw it out - and then the impossible happened. The thing moved, and not just slightly. Despite the fact that it had appeared so fixed in place that he thought it must be embedded in the cartilage, it became mobile and slid under the skin from the top of my ear down into my earlobe.

At this point, the doctor "pulled out and stitched up the wound." Sometime later, the implant returned to its original location, where it supposedly remains today. However, by a fortunate coincidence, when Strieber returned to his seat he was directly in front of me, and only about two feet away. I very carefully examined his left ear, to see if I could see any evidence of scarring, surgery, or any foreign object. I even took a close-up picture of his left ear. Nada.

# Abductology Implodes

If "Abductology" is the study of alleged UFO abductions, then it can be said that not since the sudden demise of Marxist-Leninism has any subject, real or imagined, self-destructed so suddenly and so completely as Abductology has managed to do in recent years. What has brought down Abductology so quickly? In a deadly one-two punch, a woman who was one of Jacobs' subjects is publicly accusing him of unprofessional conduct, and has recordings to back herself up. This was followed by Hopkins' ex-wife spilling the beans about his gullibility and even intellectual dishonesty.

For some time now, the matter has been simmering concerning a woman who uses the alias "Emma Woods." She was a hypnotic subject of David Jacobs from 2004 to 2007, all of which took place over the telephone. She has written, and circulated widely within UFOlogy, long and detailed accounts of her complaint against Jacobs. Assuming she can document all of her charges, Jacobs appears in a sorry light, indeed. She accuses Jacobs of telling her, during hypnosis sessions, that she suffers from Multiple Personality Disorder (MPD). She also accuses him of "planting" false memories in her of evil aliens abducting her, raping her, and even trying to kill her. She says she felt sick every time she saw the ocean because she "remembered" an alien hybrid holding her head under water. In 2006 Jacobs wrote to her in an email, "I am in a rather severe crisis with the aliens. I will be talking to them tonight about my future and what they will or will not do to me." The alien hybrids were using another woman's Instant Messenger to communicate with Jacobs (but of course she did not type the messages, the hybrids did). Apparently Jacobs reached an agreement with the aliens: he would agree to check their on-line messages frequently, and they agreed not to abduct him and implant a tracking chip. Problem solved.

"Emma Woods" is now considering legal action against Temple University, Jacobs' employer. (Jacobs has no training in medicine or hypnosis - he is a historian.) Her website is at *http://www.ufoalienabductee.com/* .

On his website *http://www.ufoabduction.com/* , Jacobs has a response to what he calls the "defamation campaign" against him. Referring to "Emma" as "Alice," Jacobs says that she appears to suffer from

"Borderline Personality Disorder," and that she has been experiencing an "emotional breakdown."

The second punch, one I was not at all expecting, comes from **Carol Rainey**, the ex-wife of Budd Hopkins, in her article "The **Priests of High Strangeness**." Upon reading "Emma's" account, she jumped into the fray:

> the trusting and vulnerable patient delivered up to Jacobs his hoped-for narrative of predatory hybrids among us— exactly what he ordered for the book he was writing. However, it's anything but a typical abductee's experience: violent sexual encounters with a human/alien hybrid; a request by the good Doctor (Ph.D. in history, non-medical) to send him her panties, unwashed, so they could be tested for alien sperm; and a proposal that she wear a chastity belt with nails across the vaginal opening, which he'd locate for her from (in Jacobs words) "a sex shop that specialized in bondage/dominance, a place that I frequented quite often."

An experienced documentary filmmaker in the medical field, Rainey soon realized that "what Hopkins and Jacobs claim as 'the powerful evidence' for alien abductions and hybrids among us is based primarily on the powerful, hypnotic repetition of their own proclamations—and the public's gullibility in believing whatever unfounded theories these star paranormal investigators punt down the field." She became increasingly skeptical of one of Hopkins' star abductees, James Mortellaro. "Several things about this case were making me increasingly uneasy. It wasn't just the pills and the pistol [that he always kept in his boot]. Or the fact that none of Jim's claims had been checked or verified. Among his more mundane statements, Jim Mortellaro had earlier told Budd that he had two Ph.D.s (Really? That's impressive, the skeptical wife thinks from behind the camera. From which universities?) and that he'd been "the Marketing Director for Hitachi" before retiring early. (Really? Was that Regional, National or International Marketing Director?)". But Budd wasn't curious. Later, Hopkins received several phone messages from individuals who called to confirm key portions of Mortellaro's story. Hopkins may have been fooled by them but Rainey wasn't: "I've spent twenty-plus years in post-production suites, with the editor or the mixer altering voices up, down, and sideways," she told her husband. "It's certainly not rocket science and Jim knows electronics. Listen, that's his syntax, that's the way he says 'very concerned' and drops his 'gs' on certain words." But instead of becoming suspicious of his "abductee," Hopkins became angry with his wife.

Rainey assisted her husband in the editing of his book on the famous "Brooklyn Bridge" abduction story of Linda "Cortile":

> It was highly dramatic, paced like a thriller— full of otherworldly treachery, forbidden love, UFOs over Manhattan, twenty-two witnesses, a heroine whose red blood cells were immortal, lusty and dangerous Secret Service agents, a Prince from afar, gifts of many fur coats, chases on foot, more forbidden love, an X-rayed alien implant, Linda's abduction into a spacecraft accompanied by an important world leader, her abduction into a spacecraft with other members of Budd's abductee support group, and her abduction into a spacecraft accompanied by a famous Mafia don. Then, later, as the story continued to unfold (long after the book's publication), Linda's presence in the lobby of the World Trade Center when the planes hit and her bloody, barefoot escape over shards of glass. Although…not all of those events reported above by Linda Cortile had been selected by Budd for inclusion in the book. I knew about them, but they weren't in the book.

The fact that the book was titled *The Brooklyn Bridge Abductions* did nothing to enhance its credibility. This story already produced a huge stink in UFOlogy during the 1990s when some UFOlogists tried to independently confirm some of Linda's wild tales, and came up with nothing (Sheaffer, 2011, p. 70). Worse yet, Hopkins "continued to tout the major significance of the case long after he knew that Linda had lied to him on multiple occasions," according to Rainey.

Another thing we learn from Rainey is that author Leslie Kean is "Budd's new protege, advisor, and all-round organizer." Now we begin to understand why Kean is so impervious to any facts that contradict her published position: she likely learned this *modus operandi* from Hopkins. Rainey notes,

> In our house, the words "debunkers" and "skeptics" were used very much in the way that devout Christians use the words "unbelievers" and "the unsaved."

# UFO Abductions

The two best-known abduction investigators, Budd Hopkins and David Jacobs, work almost exclusively alone (separately, although with extensive telephone exchanges), without supervision (and are unwilling to accept any), and without any training in medicine or psychiatry or neurology. A bit of comparative religion, anthropology, and folklore under the belt wouldn't hurt, either, in dealing with these difficult-to-interpret human experiences. They're not required to get authorization for their experimentation on human beings from an Institutional Review Board (IRB), a clearance that's required of every legitimate institutional researcher in the country. It's peer review of a proposed study using human subjects, it's strict, and researchers are required to report back to the IRB with their findings. None of this applies to UFO researchers.

In hindsight, this outcome was inevitable. Anyone who ever tried to have a rational conversation with either Hopkins or Jacobs found the two men to be extraordinarily smug, self-righteous, even pig-headed. They were correct, you were wrong, and probably stupid as well: it's as simple as that. (I never got a chance to chat with Mack, apart from a quick "hello, how are you?" in passing. The circles he moved in were far too rarefied for me to enter.) In their own circles, each was a god, more or less, and one doesn't question godlike beings. In fact, one of Hopkins' devoted fans described him literally as a "saint." There's truth in the old Biblical saying, "pride goeth before a fall." When someone smugly thinks he is invariably correct no matter how foolish his pronouncements sooner or later the Foolish Factor will grow so large that even many of his sycophants won't be able to ignore it.

With the excesses and absurdities of the Hopkins/Jacobs/Mack UFO abduction paradigm becoming increasingly obvious with each passing year, some UFOlogists have seized upon the Hill case as perhaps the 'one true' instance of claimed alien abduction. However, in view of the above, it must be painfully evident to any fair-minded person that the credibility of that classic case is essentially nil.

As might be expected, UFO abduction mania has been gradually fading as a force within UFOlogy. When abduction fever was rising, excited UFOlogists believed that it would finally deliver what UFOlogy has always wanted: validation of their personal belief in extraterrestrial visitors. But with the clear recognition by the mid 1990s the abductionists had taken their best shot and missed, UFOlogists gradually became disillusioned with abduction claims, realizing that they would ultimately fail to deliver. Thus, subsequent abduction claims failed to generate the same level of excitement. UFO abductions continue to be reported, alleged abductees continue to be hypnotized, and abductee-related support and research groups continue to operate. However, the *ennui* factor plus the serious blows recently brought upon themselves by prominent abductologists have largely quenched the once-red-hot passion for UFO abduction tales. Abductees are seen today by many UFOlogists as something marginal and/or passé and are no longer looked to as the most promising area of UFO research.

## The Demise of the Troika

On September 27, 2004, Dr. John Mack was struck by a vehicle and killed while crossing a street in London. Mack was in London to speak at a conference about T.E. Lawrence (Lawrence of Arabia), about whom he had written a Pulitzer prize-winning biography. The incident seems to have been a random tragic accident, as Mack was in a crosswalk and the driver was reported to be intoxicated. But almost immediately, some UFOlogists began to speculate that *somebody* must have wanted Mack dead, apparently because he knew too much about flying saucers. Within 24 hours of Mack's death, a note was widely circulated on the internet from somebody purporting to be "British Higher Intelligence Agent X." X says, "There's something circumstantially very odd here. Why was Mack alone in a deserted area almost totally unknown to him? For a planned hit, the Totteridge area of Mack's travels from his London dinner would be the perfect time and place. Car accident? You would have to try very hard to get hit by a car at night in this area. It is almost totally deserted at night, being right by three of the biggest reservoirs in Britain. The vast waters would have been a perfect entrance and exit route for a hit team in silent electric dinghies." The San Francisco Bohemian physicist Jack Sarfatti commented, "As Agent X says: What was John Mack doing alone late at night in such a god-forsaken place? I have recently returned from Britain myself and I would never have been alone in such a place late at night. Something stinks here." Now all that remains to be explained is why anyone would want to assemble a hit team to attack an eccentric old professor whose pronouncements about UFOs had gotten so bizarre that practically nobody took him seriously any longer.

UFO Abductions

No doubt UFO abduction claims will trickle on for a while, but it's clear that Abductology, as practiced by the Troika in its heyday, is now considered even by many UFO proponents to be an embarrassing chapter in the history of UFOs that should be forgotten as quickly as possible.

## Removing Implants, and Eyes

Some claims about "alien abduction" are too serious to be funny. Dr. Roger Leir (1934-2014) was a podiatrist who made a name for himself in UFO circles by surgically removing what he claims to be "alien implants," although he never seems to have possessed any alien artifact that can actually be examined. According to Leir's book *UFO Crash in Brazil*, "the abductors told their victim that they were going to remove his eyes and this procedure would help them further the health of the world's population. This poor man totally accepted their explanation." When he got home, "His eyes had been removed in total. He was not upset and it was only on the insistence of other family members that he consented to the opinion of a qualified opthalmologist. It will probably not come as a surprise that the doctor did not believe his story and told one of his relatives that whomever removed the eyes were (sic) a very skilled surgeon and did a magnificent job." Leir concludes, "There are cases of human mutilation that go without any publicity whatsoever. It should not surprise the reader to also learn that the United States has had cases involving some non-human intelligence that has perpetrated these sinister acts upon the U.S. population. If the truth were told to the public, it goes without saying there would be an uncontrollable panic."

Yvonne Smith is one of the few hypnotherapists still working with alleged UFO abductees on a regular basis. At a MUFON lecture in 2012, a questioner asked her if there are some people who cannot be hypnotized. She replied that anyone can be hypnotized if the hypnotist persists, although some are much harder to hypnotize than others. What she said seemed to be saying was, "they'll be good and hypnotized by the time I'm done with them!" Perhaps I am wrong. She also told about an implant that was taken by Dr. Leir from his patient Ron Noel, which according to Leir emits "deep space frequency" radio waves, and has a strong magnetic field. Smith said that this metallic implant was so hard that diamond cutting tools could not cut it. Eventually, a high-powered laser could.

At that same event, Leir's assistant Steve Colbern returned to Smith's earlier comment about Ron Noel's alleged alien implant. He added that it was made of carbon nanotubes. The aliens use nanotechnology extensively. Implants are typically 3-4 mm long, and 1 mm diameter. Leir added, from the audience, that their magnetic decay occurs over about six weeks after their removal. The isotopic ratios in the implants, according to Colbern, indicate that the objects probably originate about 30,000 light years away, near the center of the galaxy.

## Why do we have reports of 'Alien Abductions'?

Does this semi-obsession with alien abduction (and alien sexuality) represent evidence of an actual alien intrusion on our earth, or does it instead have to do with that subculture in our society enamored of science fiction and fantasy (not to mention kinky sex)?

In some instances, seemingly genuine UFO abduction accounts can be unambiguously traced back to known prosaic causes. James **Oberg** demonstrates that an elaborate 1978 sexual abduction fantasy account in **Brazil**, accepted as authentic in the UFO literature, was triggered by the sight of a Soviet **Molniya** satellite being launched into a higher orbit. Startled by his spectacular view of celestial pyrotechnics, the young man's mind then filled in the elaborate details of an encounter with a naked alien seductress whose "breasts were fuller than the breasts of a female from earth." While acknowledging that many "close encounter" or "abduction" cases follow in the wake of known satellite events, respected UFOlogists have insisted, in all apparent seriousness, that space aliens use these well-known celestial spectacles as a kind of cover, behind which they can operate unsuspected. If we accept this, the person experiencing the UFO event is spared the embarrassment of having us conclude that his or her mind wove elaborate fantasies into an event having a known prosaic cause. It is only Occam's Razor which prevents me from taking this suggestion seriously.

A "credulity explosion" has been occurring in UFOlogy since its beginning. The uniform tendency has been that the kind of stories generating excitement and attention in any given year would have been rejected by mainstream UFOlogists a few years earlier as being too outlandish. An explanation of this may lie in Martin **Kottmeyer**'s view of UFOlogy as an "**evolving system of paranoia**." He notes the disparity between "the strangeness of the claim and triviality of the proof." If UFOlogy is a "species of paranoia" then "the psychological literature on paranoia will provide insights into the dynamics of UFO belief." He links the known stages of paranoia which include "a sensitivity to external contacts," "subjective preoccupations," progressing to "recognition of an important mission such as saving the world," finally culminating in "cosmic identity" to the usual progression of claims of UFO contact.

Hypnosis is **not** a reliable method for extracting so-called "hidden memories", and its use in this manner is likely to lead to fabrication and error. Moreover, if it is suggested to a hypnotized person that fictitious events have occurred, the subject himself may come to believe this. While abductionists keep stating that supposed UFO abduction victims are an entirely normal and apparently random segment of the population, there is excellent reason to doubt this is true. A blind evaluation of nine supposed abductees by Dr. Elizabeth Slater revealed "narcissistic disturbance," "a tendency toward mildly paranoid thinking," and other disturbances. These abductees display many of the characteristics that psychologist Ernest Hartmann characterizes as "boundary deficient," people who are particularly prone to nightmares [Hartmann, 1991]. The sense of "helplessness" and "paralysis" that typify the supposed abduction experience are quite characteristic of pavor nocturnus (nightmares).

Going back a few centuries, we can find startling similarities between contemporary UFO abduction claims and the witchcraft mania. Both posit large-scale, secret operations being carried out against the human race by a realm of beings unknown to ordinary experience. Both types of incidents were supposedly difficult to uncover, requiring the assistance of an experienced investigator who, knowing what questions to ask, could 'draw out' the hidden story. It is now accepted among abductionists that many abductees have alien "familiars," who are with them on a regular basis; they are perhaps unaware that a witch's "familiar spirit" was a mainstay of the lore of witchcraft. Witches, like contemporary abductees, also had "implants": genuine witches could supposedly be distinguished by some unusual visible mark that the devil had supposedly implanted upon them during a visit. Witches, like contemporary abductees, were reported to engage in sexual rites with unworldly beings, and sexual imagery is common to the lore of both.

We could only accept the claims of "alien UFO abductions" if the evidence presented in support of such claims were more compelling than the evidence supporting current theories in physics and biology.

## The Scientific Study of Witchcraft

Joseph Glanvill (1636-1680), an early member of the Royal Society of London, performed what we would today call a "scientific study of witchcraft," using a methodology identical to that many UFOlogists now employ. He found overwhelming credible eyewitness testimony supporting its reality.:

> We have the attestation of thousands of Eye and Ear-witnesses, and those not only of the easily-deceivable vulgar only, but of wise and grave discerners; and that when no interest could oblige them to agree together in a common Lye ... for not only the melancholick and the fanciful, but the grave and the sober, whose judgments we have no reason to suspect to be tainted by their imaginations, have from their own knowledge and experience made reports of this nature. (Glanvill, 1689. See also the discussion in Sheaffer, 1998, p. 107-118).

Glanvill argued that the overwhelming consistency of the testimony among great numbers of witnesses forced one to accept the reality of the witchcraft phenomenon. He cites one case in the village of Mohra, Sweden in 1678 where some seventy witches were said to have seduced about three hundred children into black magic (numbers exceeding even those supposedly initiated into Satanism at the McMartin daycare in 1983, a tragic fiasco whose outcome is recounted in a *Wall Street Journal* article of Feb. 8, 1990). Glanvill notes that the children were examined separately by the king's Commissioners, and "all of them, except some very little ones" told stories that were highly consistent. That this consistency could have resulted from the process of interrogation itself seems not to have been considered. I suggest that whatever psychological pathology resulted in the witchcraft mania of centuries past is clearly responsible for the UFO abduction mania today.

# UFO Abductions

Occam's Razor compels us to reject explanations that are not parsimonious. How many heretofore unknown categories of "entities" would the acceptance of "UFO Abductions" create?

- Extraterrestrial intelligent visitors, of several different species (frequently hypothesized, but unsubstantiated).

- Telepathic communication (widely believed, but absolutely without scientific support).

- New laws of physics: Faster-than-light travel, anti-gravity beams, teleportation, unknown means of propulsion, etc.

- New laws of human physiology: paralysis at a distance, etc.

- New laws of genetics: not just inter-species mating, but inter-planetary.

Since this sort of mad multiplication is exactly what Occam's Razor tells us we are *not* supposed to do, the only tenable hypothesis for" UFO abductions" is the *null hypothesis* – that nothing is going on!

# 6. UFO CONSPIRACIES

Given the near-universal belief among "Science Fiction" UFOlogists that UFO crashes, secret programs, and even alien captures have taken place, it follows that there must exist conspiracies of gigantic scale with vast resources to successfully conceal UFO secrets from the world at large. Furthermore, all major nations must be colluding in this effort; it would do no good for the U.S. to hide its UFO secrets from the world if France, or China, or Brazil also had secret information about UFO crashes, and decided to release it.

There is a widespread belief in an alleged secret U.S. government group known as MJ-12 (or **Majestic-12**), whose job is to investigate UFO crashes and also arrange their cover-up. UFOlogists William L. **Moore** and Jaime **Shandera** announced in 1987 that they had anonymously received copies of supposed government documents purporting to show the activities of this secret UFO crash/retrieval organization. When the FBI investigated these "documents," they quickly concluded that they were "completely bogus." Other problems in the documents were soon noted. For example, one document was typed on a typewriter model that was not manufactured until fifteen years after the date on the document. One MJ-12 document was reportedly found in a National Archives repository in Maryland. However, unlike authentic archive documents, that one has no archive registration number, and appears to have been smuggled into the archive in order for it to be later "found." Many UFO proponents strongly defend the authenticity of these "leaked" MJ-12 documents, but no proof of their authenticity has ever surfaced.

Almost seven years after the appearance of the original series of MJ-12 documents, a second series of documents supposedly began to be leaked to UFOlogist Timothy S. Cooper in California, starting in 1999. These new supposed documents far outnumber those that Moore and Shandera claim to have received. These newer MJ-12 papers are even less credible than the original ones, and even UFOlogists who supported the original MJ-12 documents like Stanton Friedman and Timothy Good have expressed doubts about the second series. Dr. Robert M. Wood and his son Ryan S. Wood are the principal promoters of the "Majestic Documents" today via their Web site majesticdocuments.com, documentaries, conferences, etc. Philip J. Klass noted that Robert Wood admitted that there were flaws in the documents. But Wood claimed that these anomalies "tend to indicate authenticity.... [Document] hoaxers generally try to make sure they are perfect."

Some hypothesize that NASA is involved in a giant conspiracy to hide UFO data uncovered during its various space flights. Rumors of astronaut UFO sightings abound, supported by misquotations and even outright fabrications. Comments from astronauts concerning sightings of not-then-identified space debris were taken out of context to make it sound as if they saw alien spacecraft. While a few astronauts have been believers in UFO claims (most notably Edgar Mitchell and the late Gordon Cooper), not one astronaut claims to have seen any non-earthly technology while on any spaceflight. During shuttle missions while the astronauts are sleeping, NASA often made real-time video available of Earth from the orbiter's cameras, which was shown by some cable-TV services on the NASA channel. Some UFOlogists, convinced that there are secret goings-on concerning UFOs and NASA, recorded many hours of this uneventful video. Later, they scrutinize the recordings, looking for little dots or blips that to them represent alien spacecraft. Tiny pieces of ice or other orbital debris, sometimes kicked around by exhaust from the shuttle's attitude control thrusters, were trumpeted as proof of aliens cavorting about while watching our space missions, a secret said to be kept hidden by NASA. **James Oberg** has written many articles refuting claims of **NASA UFO conspiracies**.

## Richard C. Hoagland - Space Conspiracy Guru

Author Richard C. Hoagland has become famous by promoting claims of many weird space-related claims and conspiracies, mostly involving NASA. Over the years he has claimed that NASA has been covering up knowledge of a face on mars, large alien artifacts on the Moon, anti-gravity forces, and civilizations on the moons of Jupiter and Saturn. His Web site, enterprisemission.com, is filled with claims of space conspiracies and "hyperdimensional physics," which apparently goes far beyond anything known to ordinary physicists.

NASA's highly-successful Cassini mission to Saturn has returned a wealth of scientific data, especially the Huygens probe that landed on Saturn's moon Titan, the first such landing on a planetary satellite other than our own Moon. However Richard **Hoagland** claimed in 2005 to have made dramatic discoveries from its photos, not of Titan, but instead of Saturn's moon Iapetus, an unusual body having one relatively dark hemisphere and one lighter.

> In our opinion, Cassini's discovery of "the **Great Wall of Iapetus**" now forces serious reconsideration of a range of staggering possibilities … that some will most *certainly* find upsetting: it could really *be* a "wall" … a vast, planet spanning, *artificial* construct!!… There is no viable geological model to explain a *sixty thousand-foot-high, sixty thousand-foot-wide, four million-foot-long* "wall" … spanning *an entire planetary hemisphere* - let alone, located in the *precise plane* of its equator!

Iapetus does indeed have a strange raised 'ring' across part of its equator, but it appears to be an artifact of that moon's formation under Saturn's powerful tidal forces. Hoagland has further discovered that some of Iapetus' craters appear to be somewhat "square", and are allegedly lined up along north-south, east-west lines:

> Clearly these are NOT random, "square craters" -- but remarkable, highly ordered evidence of sophisticated, aligned, repeating *architectural* relief! … The impression of a vast set of extremely ancient *ruins* – most now without roofs, but with ample surviving walls – covered both by "snow" … and whatever the "brown stuff" is … is unavoidable.

Stressing the near-impossibility of these structures being built by living beings on the surface of a cold, airless, waterless world, Hoagland leaps to a conclusion that presupposes an even greater engineering impossibility: "what if Iapetus is *not* a natural satellite at all … but a 900-mile wide *spacecraft* - an *artificial* 'moon?!'." He suggests that Iapetus was assembled millions of years ago by some alien intelligence using the principles of Buckminster Fuller's geodesic domes, and that many of the craters are in fact "deformed *hexagons.*" This brief summary cannot possibly do justice to the zaniness of Hoagland's "hollow Iapetus" theory – you need to read the original on his website.

It is telling that Hoagland does not discuss the problem that his "hollow Iapetus" theory poses for accounting for that moon's measured mean density of about 1.21 grams per cubic centimeter, less than our own moon's, but greater than that of Saturn. This is perplexing, since he begins the piece arguing that Iapetus' relatively low density and slow rotation means that the centrifugal force at the equator would be extremely small. It's too bad that Hoagland didn't follow through and give us any calculations designed to show how it would be possible for an essentially hollow sphere to have a mean density greater than that of water. He must have realized the fatal flaw this calculation would pose for his wild assertion, which is why he avoids the subject.

When NASA's highly successful **New Horizons** mission returned detailed close-up photos of **Pluto** in 2015, **Hoagland** was quick to claim that they showed evidence of alien life. He also claimed that the close-up photos of **Ceres** by DAWN that same year revealed alien structures on that dwarf planet, including a three mile high "pyramid."

## All High Weirdness, All Night Long

Hoagland became one of Art Bell's perennial favorite guests on the all-night high-weirdness radio show *Coast to Coast AM*, now regularly hosted by George Noory. Hoagland generated tremendous excitement among listeners with his claim that an alien space probe was due to land near Turret Mountain in central in Arizona December 7, 1998. It began with widespread reports made on the internet concerning a supposed ETI radio signal that was allegedly being received from the star EQ Pegasi. According to *The SETI Institute*, this appears to have started as a deliberate prank by someone impersonating a certain SETI researcher. Hoagland somehow determined that the supposed SETI signal portended the imminent landing of an extraterrestrial probe. He told Art Bell's listeners that he was receiving information about the forthcoming landing from "intelligence community" sources, as well as from "speech reversals" (playing backwards the tapes of officials' speeches, which supposedly reveals what they are *really* saying).

Peter Gersten was then the head of CAUS - *Citizens against UFO Secrecy* , a group that has been filing UFO-related Freedom of Information Act requests for many years. Gersten sent out a cry for assistance to his members to be ready for the probe's potential arrival, since "CAUS believes we must assume that Dick Hoagland's information, based upon his experience, expertise and intuition and corroborated by his calculations and Pentagon sources, is reliable and accurate." Gersten reported that "the response was overwhelming." Pilots of small planes were asked to be on the ready in Sedona, Arizona. Hoagland, however, urged caution since "If there is a secret military tunneling operation or an actual 'ET event' is to take place, it may not be safe for you to approach the area," and he speculated that the area would be placed under martial law. Of course, December 7 came and went without any unusual occurrences. So far as is known, no unusual military activity took place in the area, and nobody found any ET probe. There was, however, what Hoagland termed "evil weather" in Phoenix on that date, which he speculated "may have *actually been induced* as part of a 'Hyperdimensional Physics' experiment."

The April, 1998 issue of CAUS' newsletter stated that "CAUS, in conjunction with Trans Lunar Research, is working on a project to send a privately funded rocket to the Moon to transmit live photos from the Sinus Medii region to verify the alleged artificial structures located there." Some UFOlogists claim to discern, in NASA photos of that region, "lunar bridges spanning vast chasms, towering structures protruding from large craters - and UFOs rising from the surface of the moon". If you've never heard of Trans Lunar Research, that isn't surprising, unless you're a fan of *Coast to Coast AM*, where they've been featured several times. The group says they've been launching small rockets in the Mojave Desert, and they certainly have big plans: their aim was to establish a 50-member lunar base, with the first inhabitants to land in 2005. While these plans sound impressive, the group itself would be much more credible if it had nothing whatsoever to do with Art Bell, or with UFOs.

CAUS estimated the cost of the unmanned lunar reconnaissance at $12 million. For that sum, they hope to be able to "verify, through photographic evidence, the existence or non-existence of alleged ancient artificial artifacts in the Sinus Medii region." Another goal of the mission was "to confirm previous moon landings," about which many people have their doubts, especially those who listen to Art Bell. What CAUS doesn't realize is that even if they succeeded in photographing Apollo lunar landing sites, Art Bell would soon feature guests claiming that *their* photos had been faked. In 2009 NASA's Lunar Reconnaissance Orbiter finally did succeed in clearly photographing some of the Apollo moon landing sites, followed by even higher resolution photos later. But the moon conspiracy believers are still not convinced.

CAUS was also keenly interested in the Cydonia region of Mars, where many UFOlogists believe that a supposed stone "face" and alleged "pyramids" were recently re-imaged during NASA's Global Surveyor mission. CAUS notes that Dr. Mark Carlotto, who formerly was suggesting that the region called "the City" may contain artificial structures, after studying the recent images, now says that the features appear to be natural formations. But Richard Hoagland isn't giving up so easily. Hoagland insists that the new photos reveal that the entire Cydonia region is made of glass. Hoagland further suggests on his website that a certain mottled image may actually reveal a "Levittown on Mars" (referring to a post-war US housing development noted for its nearly-identical, low-cost houses). Hoagland also finds in the new Martian images a "Castle of Barsoom."

Linked into Hoagland's web page was a group called *The Millennium Group*. They are raising the alarm about supposed "new solar system members" being detected by the SOHO satellite. When contact with SOHO was temporarily lost in June 1998, there was speculation by UFOlogists that the satellite had spotted something that it *wasn't supposed* to see, and may have been a victim of "alien attacks." One website hosted by "Citizens Against the New World Order" called "CyberSpace Orbit" was promoting bizarre explanations for SOHO's demise. It claimed that two weeks earlier a kind of solar flare "seemed to reach out and blind" SOHO's cameras, and also that in another SOHO photo "two strange objects" appeared near the sun. Suddenly, the CyberSpace Orbit website 'disappeared' from the internet. Speculation was rife that the website was shut down because "Perhaps too many people are seeing things someone doesn't want them to see." But if the New World Order is censoring websites, they are not very good at it, because the supposedly banned information could soon be viewed in a new location.

Once SOHO recovered fully from its "alien attack," it was again sending images, again including blurry objects representing solar phenomena and optical artifacts. One of the supposed new solar system objects has been dubbed "Orca" by The Millennium Group because of the resemblance of its blurry, formless shape to a whale. "The LASCO C3 photo shown above was the very last image before the SOHO/LASCO satellite experienced a 90 day shut down. We are now more confident than ever that the interruption in service of the SOHO/LASCO satellite was done on purpose by the US military/intelligence community to prevent the public from seeing any more of the "Orca", or the other new members of our inner Solar System. We also have reason to believe that the DIA, NSC, and/or NSA are working independent of any congressional or administrative authority."

The above, however, gives us only a small hint at the full richness of misinformation to be found by listening to *Coast to Coast AM*. Among the other gems that have been gleaned by regular listeners:

• A caller in northwest Washington told Bell that he had mysteriously suffered an hour and a half of "missing time." Bell suggested that he be "regressed" by a hypnotist immediately, to check for possible alien abduction. The caller said that he had talked to others who had experienced the same thing at the same time, causing Bell to wonder if perhaps time for everybody might have "slipped." A week later, Bell was claiming that hundreds of other people in Seattle had sent him faxes saying that they, too, had experienced "missing time" at this same moment.

•Dr. Jonathan Reed says he ran across an alien in the woods, causing his dog to implode. In self-defense he struck the creature in the head, and it fell. Believing it to be dead, he wrapped it in a blanket to carry it home, and placed it in his freezer. Three and a half days later, he opened the freezer, and the not-so-dead alien let out a piercing, unearthly scream. Unfortunately, the Men In Black must have gotten word of this amazing discovery, because soon his house was broken into, the intruders removing not only the alien, but the freezer as well. As proof of his story, Reed offered photos of the dead alien, the hovering craft, and even the recorded sound of the creature's scream.

## A Secret Space Program

According to UFOlogist Linda Moulton Howe, there are huge honking machines orbiting the earth. She has written extensively about a "Secret American Military Space Program." She interviews those who claim to have knowledge of a secret military program far advanced beyond the public one we know about, run largely from secret underground military bases. The proof? Decades-old newspaper clippings, describing ambitious space projects that never got off the ground. Aah, but they did – and promptly went underground! This conspiracy apparently also involves the Masons, because one high-ranking NASA Apollo manager was a Mason of the 33rd degree, as well as esoteric ancient Egyptian wisdom, because an Egyptian geologist worked on lunar science for the Apollo program. (How can anyone dismiss solid evidence like that?)

Richard Dolan is another contemporary UFO author and lecturer who has made outrageous claims about government UFO conspiracies. He told the International UFO Congress in 2013 that there is a "secret space program," whose mission is to investigate supposed "anomalies" on the moon, and Mars.

## Extraterrestrial Technology

Another major purported conspiracy centers around the claims of reverse engineering of alien technology made in *The Day after Roswell* (1997) by the late Col. Philip J. Corso. According to Corso, a great deal of today's familiar technology, including integrated circuits, fiber optics, and lasers, were not actually invented by earthlings but were reverse-engineered from technology found in the alleged Roswell saucer crash. Corso also claimed that he alone was able to understand this alien technology, after some of the nation's top scientists had tried and failed. Corso's claims have been extensively investigated and debunked by pro-UFOlogists Brad Sparks, John Alexander, Jacques Vallee, and others but continue nonetheless to enjoy widespread acceptance, in spite of being entirely without proof.

Starting in the late 1990s, a dynamic young internet entrepreneur in California's Silicon Valley became a serious contender for making the most outrageous UFO claims. Joseph Firmage (who we met in chapter 5), then just 28, was founder and CEO of the highly successful U.S. Web Corporation, founded just four years earlier but then worth $2.8 billion, more or less, depending on whether it was a day when the NASDAQ is up or down. However, Firmage said in an interview that his role at the company had diminished from CEO to "founder and chief strategist" to "the guy at the end of the hall who believes in aliens." Like Saul of Tarsus,

Firmage had his life altered forever by a vision, only unlike Saul his vision didn't occur on the road to anyplace, but instead while lying in his bed. "A remarkable being, clothed in brilliant white light, appeared hovering over my bed in my room. Out of him emerged an electric blue sphere, just smaller than a basketball, which was swirling with what looks like electrical arcs. It left his body, floated down, and entered me." After a brief conversation with the being, he was convinced that aliens were real. Had Firmage known more about psychology, he would not have embarrassed himself with this account: This is a classic "hypnopompic hallucination", well-known in psychology, and frequently cited in the serious literature dealing with supposed "UFO abductions." (for example, see "Abduction by Aliens or Sleep Paralysis?" by Susan Blackmore, *Skeptical Inquirer,* May-June 1998). Halfway between the dream state and waking, the mind confuses the input it receives from both. Firmage wrote an on-line book about his beliefs and experiences, modestly titled *The Truth*.

Firmage used $3 million of his own money to found the International Space Sciences Organization in Santa Clara. He soon was accepting the most outlandish UFO claims making the rounds, the whole enchilada, MJ-12 and the like. Firmage, too, preached the gospel that most of our newest high-technology wonders - for example, fiber optics - were reverse-engineered from objects found in a crashed saucer in Roswell, surely among the most absurd and ironic statements ever made by a Silicon Valley insider. Dr. Narinder Kapany, who developed the first fiber optics in the 1950s, exclaimed to the San Jose *Metro* "that makes me a spaceman!" He explained that his interest in bending light around corners began when he was a student in India during the 1940s.

## Rendle-sham Shenanigans in the UK

The supposed Rendlesham Forest UFO landing case (sometimes referred to as "the British Roswell") involved the supposed landing (or at least Close Encounter - the story is inconsistent) of a UFO in Rendlesham Forest, Sussex, UK in December, 1980. Just like the story of the fish that got away gets bigger with each telling, the more time that passes, the more exciting the Rendle-sham case becomes. Practically each year, one of the supposed witnesses invents a new and dramatic claim to 'prove' that the case is real.

# UFO Conspiracies

British skeptic **Ian Ridpath** has long stayed on top of **Rendlesham**. Here is his summary of it:

Although the overall case is complex, the main aspects can be summarized as follows:

1. Security guards saw bright lights apparently descending into Rendlesham Forest around 3 a.m on 1980 December 26. A bright fireball burned up over southern England at the same time.

2. The guards went out into the forest and saw a flashing light between the trees, which they followed until they realized it was coming from a lighthouse (Orford Ness).

3. After daybreak, indentations in the ground and marks on the trees were found in a clearing. Local police and a forester identified these as rabbit scrapings and cuts made by foresters.

4. Two nights later the deputy base commander, Lt Col Charles Halt, investigated the area. He took radiation readings, which were background levels. He also saw a flashing light in the direction of Orford Ness but was unable to identify it.

5. Col Halt reported seeing starlike objects that twinkled and hovered for hours, like stars. The brightest of these, which at times appeared to send down beams of light, was in the direction of Sirius, the brightest star in the sky.

At its most basic, the case comes down to the misinterpretation of a series of nocturnal lights – a fireball, a lighthouse, and some stars. Such misidentifications are standard fare for UFOlogy. It is only the concatenation of three different stimuli that makes it exceptional.

The BBC reported on July 13, 2015 that Col Halt is now claiming,

"I have confirmation that (Bentwaters radar operators)... saw the object go across their 60 mile (96km) scope in two or three seconds, thousands of miles an hour, he came back across their scope again, stopped near the water tower, they watched it and observed it go into the forest where we were," said Col Halt.

"At Wattisham, they picked up what they called a 'bogie' and lost it near Rendlesham Forest.

"Whatever was there was clearly under intelligent control."

Halt does not name the supposed radar operators, and does not say how he obtained this information. He claimed that the operators said nothing about this until after their retirement, for fear of being "decertified" for reporting a UFO. Even if this unlikely claim were true, it does not correspond to what the supposed witnesses are reporting. A UFO allegedly whizzing by at thousands of miles an hour does not match the UFO(s) allegedly seen hovering for hours above the forest, and even landing there.

The Daily Express of London reported on July 14, 2015 that Halt said:

• Rendlesham Forest was mobbed with US military personnel "hunting UFOs" at the time.

• A "UFO exploded " before his eyes and another "shot down a laser beam" from 3,000 feet above

• The UK Ministry of Defence (MoD) later hid documents relating to Rendlesham

• US personnel who "lost 40 minutes" during the sighting have been denied access to medical files

Speaking at a UFO conference in Woodbridge, Suffolk, he said: "There is no doubt in my mind we are not alone and there are some people (in power) who know this, but even Mr (Barack) Obama won't get through to them."

This is not the first time that Halt has 'jazzed up' his account of the Rendle-sham incidents. Ridpath describes Halt's "iffy affidavit," written in 2010, as a "disastrous attempt to rewrite the facts of the case," suggesting that "this product of his 30-year-old memory differs so substantively from what he said and wrote at the time that it would be destroyed in a court of law."

What are other "top witnesses" from Rendle-sham reporting?

# UFO Conspiracies

Airman Larry Warren claimed to have seen a light in the forest that "blew up," then re-assembled itself, and alien beings came scampering out of it. He says they resembled "children in snowsuits." By his account, many other Air Force personnel saw these creatures, but nobody else has reported them.

Another supposed witness, John Burroughs, has implied while supposedly under hypnosis that he and Sgt. Penniston were abducted by beings onto the UFO for about 40 minutes, and brought back to a different place. Supposedly base personnel saw them being lifted up to the object, and worried that they would never be returned. Currently Burroughs is charging that his inability to obtain medical care from the Veterans Administration is tied in with his alleged UFO encounter, and that his medical records are "classified." Others see him as just one among many victims of the ongoing Veterans Administration scandal, fraud and false documents used to conceal the fact of thousands of veterans being denied the medical care that is their right.

Sgt. Jim Penniston (ret.), however, relates a completely different UFO yarn, in spite of supposedly sharing in Burrough's UFO abduction. He claims to have spent the 40 minutes examining and touching a landed UFO, and to have received a telepathic message from it in the form of a "binary code," which he wrote down in real-time. However, he did not tell anyone about this for over twenty years. Ridpath points out that Penniston wrote down the wrong date and time for the event in his notebook .

Penniston now says that the binary data from the Rendle-sham UFO was sent by Time Travelers, which makes sense since aliens would hardly be expected to encode their messages using ASCII, the American Standard Code for Information Interchange. But time travelers from a future earth might possibly still be using ASCII 6,000 years from now (and probably Microsoft Windows as well). When Penniston was interviewed on a 2011 podcast by Angelia Joiner, his story got all tangled up under her questioning. Penniston got confused whether he knew what the code meant as it was being transmitted. He finally decided that he knew what six pages of it meant, but he thought that the rest of it "didn't mean nothin." Nor did Penniston's alien ASCII codes end in 1980. Several of Penniston's associates relate how he has been receiving more such codes as recently as 2011. In any case, Penniston's magic notebook, which is like a cornucopia producing marvelous revelations each time it is opened, has open-able rings in the center, allowing pages to be taken in or out.

What does Penniston's supposed "binary code" supposedly mean? Nick Pope writes,

> Many people have attempted to translate the Rendlesham Code. The best-known suggestion so far is "Exploration of humanity. Continuous for planetary advance. Eyes of your eyes. Origin year 8100". Also included is a list of geographical co-ordinates which seem to match up with a number of mysterious and ancient sites such as the Great Pyramid in Egypt and the Nazca Lines in Peru. Another co-ordinate corresponds with the fabled location of the 'lost continent' of Hy-Brasil, a supposed sunken island in the Atlantic, off the West coast of Ireland, sometimes dubbed the "Celtic Atlantis."

So it's supposedly a binary message from the Celtic Atlantis in the year 8100. If you can describe any of these "witnesses" as "credible," then you are much better at "believing things" than me.

Finally, the Rendlesham Forest now has its own official "UFO Trail," with a map and an explanation helpfully posted on a sign (http://goo.gl/fXXZL9). A UFO encounter is a terrible thing to waste. The British UFOlogist Nigel Watson has suggested that aliens may be responsible for dogs developing a "mystery illness" after having walked on that trail.

Penniston may have started a trend. On June 29, 2015 at 1:28 AM, a man (identified only as C.J.) and his wife were driving near Wadley, Georgia, when a "Large metallic object with fluid oil slick body spurting sparks from the rear flies dirctly over Our truck" (sic). But satellite orbit guru Ted Molczan found that the time of the event matched almost perfectly with the known re-entry of 1973-084D / 6939, the final rocket stage of the launch of Cosmos 606. This event was widely witnessed, giving rise to 169 reports in the database of the American Meteor Society. The witness, however, "states unambiguously that he believes that what he witnessed with his family was not a re-entering satellite."

In fact, according to the Crop Circles Research Foundation, this witness (known as C.J.), was unconsciously compelled to write down symbols consisting of lines and squares from left to right, on the back of the Motel 6 full page receipt....It has been independently confirmed both by myself and another researcher (Dr. Horace Drew) that C.J.'s symbols are in fact a message written in binary code that can be translated via a standard ASCII table."

Exactly as Penniston claimed to have done! C.J.'s message has been translated to mean:

Continuous protection of humanity 49.27n, 11.5e. Expose Hidden Knowledge to ALL citizens. Advancement imperative for planetary survival. Beware of Orion 1350.3 and Z Ruticuli 39.170. Avoid [signal] messages sent.

## 2009: NASA Tries to Bomb Star Visitors

NASA may have most people convinced that its purpose in crashing the Lunar Crater Observation and Sensing Satellite (LCROSS) satellite into the moon on Oct. 9, 2009 is to look for ice in a permanently-shaded crater near the moon's south pole. But the well-known UFO expert Dr. Richard Boylan of Sacramento, California, isn't fooled – he knows that it's "a Cabal project to annihilate a Star Visitor colony living in a crater near the Moon's South Pole." Boylan, a former psychologist who lost his license over allegedly improper behavior, is a board member of a group called *The Academy of Clinical Close Encounter Therapists*. Boylan not only works with those who believe themselves to be victims of UFO abduction, but also detects and counsels "Star Kids" and adult "Star Seeds," people who believe themselves to have special "advanced abilities" and a special alien "mission" on earth. His website http://www.drboylan.com/ helpfully provides a checklist for those who believe that they or their children may be Star-special. Answer "yes" to twenty or more of the questions, and your child is "absolutely a Star Kid."

Boylan explains that "The Cabal within NASA know that there is a colony of Star Visitors living within Cabeus A Crater. The Cabal's secret objective is to use the LCROSS and attached rocket stage to obliterate the Star Visitor settlement residing within that crater...I note that the Cabal is indeed engaged in unlawful war crimes and attempting to position the United States, and by extension, all Earth nations, in an act of war against star civilizations. Since this is not a true act of the United States Government but a rogue act by Cabal infiltrators within NASA, then the official government of the United States, and by extension the United Nations, would repudiate this action as unlawful once its true intent becomes known."

To try to head off this disaster, Boylan attempted to send a message through unspecified special channels to warn President Obama, and also Vice President Biden, "who normally oversees the government's Star Visitors programs." Unfortunately, the message did not get through because it was intercepted by Secretary of State Hillary Clinton, who is a "Cabal asset." So Boylan sent a telepathic message to Star Nations High Council, asking if they would like him to organize a "Joint Psychic Exercise to redirect LCROSS and Centaur rocket away from the Moon." Receiving a reply in the affirmative, Boylan announced the following:

> 20 days from now we will engage (along with Star Nations) in a Joint Psychic Exercise to divert the LCROSS space probe and accompanying Centaur rocket away from crashing into the Star Visitors lunar colony within Cabeus A Crater. That Joint Psychic Exercise will take place simultaneously globally on October 8, [the day before expected impact].

Boylan calls this the "Joint Psychic Exercise to deflect and disintegrate LCROSS space probe and its Centaur booster rocket," and gives the hour in each time zone for his followers to perform their feats of psychic action-at-a-distance.

However, a week before impact, NASA changed its mind about which crater to impact. NASA scientists decided that the main crater Cabeas was more likely to contain significant amounts of water, and they directed LCROSS and its Centaur rocket to the new target. So probably the energy from the future Joint Psychic Exercise went back in time, and caused NASA to direct its impact away from the Star Nations visitors. Or else Boylan's urgent message finally got through to Star-Visitor-Overseer Joe Biden, who averted an interplanetary war by moving the LCROSS target. However, Boylan himself seems unaware of the re-targeting, or at least did not mention it on his website.

Precisely at the predicted time, the Centaur rocket, followed quickly by LCROSS itself, both of them undeflected and undisintegrated, slammed into the lunar crater Cabeas at a speed of about 40 km/second. Nonetheless, Boylan proclaimed the exercise a "success," claiming that the probe and the rocket were "deflected" from the Moon, and "disintegrated in space." Boylan explained how he projected himself astrally through time and space and (still apparently unaware of the probes' retargeting), "went out psychically to LCROSS and Centaur booster as they were streaming towards the Moon. Next I enwrapped LCROSS in a telekinetic force and redirected it onto a course to the left so it was aiming towards one Moon-diameter's width left of the Moon's left side. Then the same was done with the Centaur booster rocket." But merely to deflect the objects was not enough: "I engaged first one, then the other, with strong dissolution energy to unbind the Strong Force bonds holding their atoms together as molecules. [That, however, is an electromagnetic bond, not a nuclear one.] Moving from top to bottom, I un-did the Strong Force bonds, causing the component materials of these space vehicles to come apart at the molecular level. This process also safely dismantled the advanced munitions which were secretly aboard these space vehicles... This was confirmed this morning by Star Nations, whose members were also at work on these two space vehicles during our JPE, to assure thorough deflection and disintegration. Thus the star folks lunar colony within Cabeus A Crater is safe from overhead bombardment." Perhaps this explains why no ground-based telescopes observed any dust ejected from the collision?

## Jacques Vallee and the "Pentacle Memorandum"

Jacques Vallee's diaries from 1957-69 were published as Volume I of *Forbidden Science*. It tells the story of his life from childhood in France, his education and early career, his developing interest in UFO reports, etc. The book is very literate, very personal, and in places even poetic. It is a good read.

Perhaps the most significant new issue discussed in this book is the matter of the so-called "Pentacle Memorandum." In June of 1967 while Hynek was away on vacation in Canada, Vallee went over to Hynek's empty house (with permission) to organize and sort Hynek's disorganized UFO-related files. He found one document, a two-page typed memo, that he believed to be extremely significant. Dated January 9, 1953, it was stamped "SECRET - Security Information" in red ink. Vallee refered to it as the "Pentacle Memorandum," in order to not identify the author. "**Pentacle**" was later revealed to be H.C. Cross of Battelle Memorial Institute, Battelle's liaison with the Air Force for Blue Book-related matters. A good account of the memorandum's discovery, Vallee's claims about it, and the text of the memorandum itself, was written by **Philip Coppens**.

## Vallee 'discovers' a Secret Large-Scale UFO Research Program

The Memorandum begins,

> This letter concerns a preliminary recommendation to ATIC on future methods of handling the problem of unidentified aerial objects. This recommendation is based on our experience to date in analyzing several thousands of reports on this subject.

ATIC was the Air Force's Air Technical Intelligence Center. Vallee writes, "This opening paragraph clearly establishes the fact that prior to the top-level 1953 [CIA] Robertson Panel meeting somebody had actually analyzed thousands of UFO cases on behalf of the United States government" (Vallee 1996, p. 284. Emphasis in original). This is supposed to reveal the existence of a huge and secret UFO investigative program, other than Blue Book, somewhere within the U.S. government.

The plot thickens when Vallee finds out a short time later that "Pentacle must indeed have worked at Battelle [Memorial Institute]." (Vallee 1996, p. 294). Gasp - you mean that in January, 1953 there was someone working at Battelle who had analyzed "thousands" of UFO reports for the U.S. government?? And this was a secret program??

**Earth to Vallee:** The Battelle Memorial Institute began working on *Special Report #14* for Project Blue Book in March, 1952. (I don't know what happened to Special Reports #1 through #13. One never hears about them.) This was a statistical analysis of UFO reports in the Blue Book files, the first to use newfangled computers and punched cards. A total of 3200 cases were analyzed. The report was completed and published in 1954. *Blue Book Special Report #14* is well-known to UFOlogists. In fact, Stanton Friedman hardly ever stops talking about it.

Of course there were people working at Battelle in January, 1953 who had analyzed "thousands" of UFO reports for the U.S. government. They were working on *Blue Book Special Report #14*. They finished the following year.

Congratulations, Jacques! You've found indisputable proof of the existence of the team writing *Blue Book Special Report #14*! Which has never before been doubted. Just ask Stanton Friedman.

## Vallee 'uncovers' the Manipulators Manipulating the CIA's Robertson Panel

One of the most controversial sentences in the Pentacle memorandum reads,

> Since a meeting of the [CIA's Robertson] panel is now definitely scheduled we feel that agreement between Project Stork and ATIC should be reached as to what can and what cannot be discussed at the meeting in Washington on January 14-16 concerning our preliminary recommendation to ATIC.

According to Hynek, White Stork was a former Air Force project name encompassing the Blue Book project. Vallee suggests that Project Stork was keeping the soon-to-meet Robertson Panel in the dark and would decide what they would be allowed to learn. (The Robertson Panel was a then-secret meeting in January, 1953 organized by the CIA bringing together its best experts to discuss what the "saucers" really are, and what to say about them to the public.) Vallee writes that the memorandum seemed to dictate "a key determinant in what the panel could discuss – and what not, i.e. what would be kept away from the panel. By preselecting the evidence, the conclusion the scientists would reach could thus be known in advance."

Vallee should read that sentence more carefully. It does not talk about UFO sightings or evidence. He is interpreting that sentence as if it said "agreement between Project Stork and ATIC should be reached as to what can and what cannot be discussed at the meeting in Washington on January 14-16." But the sentence does not end there. It continues with "concerning our preliminary recommendation to ATIC." In other words, not to decide what subjects are off-limits to discussion by the Robertson panel, but to decide what to tell that panel about plans involving Battelle and ATIC. Or, in plain English, "How much should we tell the Robertson panel about what we've been proposing to ATIC?" I realize that English is not Vallee's first language, however his mastery of English seems to me to be so complete that I am surprised to see him misreading that sentence so badly.

## Vallee finds 'evidence' of a huge covert UFO deception project

In this passage Vallee finds evidence of a huge and alarming military-sponsored project intended to deceive the public about UFOs:

> we recommend that one or two of theses areas be set up as experimental areas. This area, or areas, should have observation posts with complete visual skywatch, with radar and photographic coverage, plus all other instruments necessary or helpful in obtaining positive and reliable data on everything in the air over the area. A very complete record of the weather should also be kept during the time of the experiment. Coverage should be so complete that any object in the air could be tracked, and information as to its altitude, velocity, size, shape, color, time of day, etc. could be recorded. All balloon releases or known balloon paths, aircraft flights, and flights of rockets in the test area should be known to those in charge of the experiment. Many different types of aerial activity should be secretly and purposefully scheduled within the area.

About it, Vallee writes, "the Pentacle proposal goes far beyond anything mentioned before. It daringly states that 'many different types of aerial activity should be secretly and purposefully scheduled within the area'. It is difficult to be more clear. We are not talking simply about setting up observing stations and cameras. We are talking about large-scale, covert simulation of UFO waves under military control."

Pentacle's proposal seems to be this: Let's identify an area where people are making a large number of UFO reports. Let's set up an extensive monitoring system so that we know everything flying in or out of that area. Then we'll try a controlled experiment: we will cause the people to see balloons, unusual aircraft activity, etc., and then monitor UFO reports we get from that area. We will see how a known stimulus is reported as an unknown object, and thereby better understand the UFO reports we receive.

This sounds like an excellent idea from a standpoint of science, although from a standpoint of law or ethics it may not pass muster. It also sounds rather expensive, and not easy to keep under wraps, which would defeat its purpose. Interestingly, as a result of several passive (not active) experiments of this kind, UFOlogist Allan Hendry, one of Hynek's chief investigators during the 1970s, became far more skeptical about eyewitness reports. As detailed in his book, **The UFO Handbook**, **Hendry** examined the reports being received originating from a known stimulus (advertising aircraft, balloons, etc.) and found many of them so wildly in error that he cautioned against taking such reports at face value. To other UFOlogists, Hendry seemed to be guilty of horrid blasphemy (even though he believed some UFO reports to be unexplainable), and they began to denounce him. Understandably embittered, Hendry withdrew from UFOlogy about 1981, and has refused to discuss it since then.

A frequent theme in *Forbidden Science* is Vallee's commentary about the rigidity of bureaucracy, in government and in science, in France and in the U.S. He gives one example after another of seemingly good proposals being rejected or even ignored by a bureaucracy unwilling to accept change. What is surprising is that here Vallee, of all people, seems to be confusing a proposal with a project. He must surely realize that, merely because Pentacle is proposing some grand and new UFO investigative project, the odds of that proposal being actually implemented by a rigid Air Force bureaucracy (which clearly had little enthusiasm for UFO investigation) were slim to none. This passage does not in any way establish that such a controlled experiment involving UFO stimuli was ever carried out.

## Vallee and Hynek's Bizarre Investigations

Surprisingly, Vallee writes quite seriously about matters such as:

- **Rosicrucianism:** They claim that their 'Ancient Wisdom' is thousands of years old, but there is absolutely no proof of that. "I find their documents to be an interesting spiritual complement to my scientific training. Every month I receive a set of course material through the mail. It includes both theoretical reading and instructions for simple rituals, promising insight into higher realities" (Vallee, p. 39). He later explains that the Rosicrucian order he belongs to is "AMORC, which is headquartered in San Jose." If you've ever been to that Rosicrucian Museum with the awesome mummies interspersed with cheesy claims of ancient mysteries, that's the group he was talking about. When I was a child, they used to regularly have ads on the back page of comic books. Hynek was also interested in Rosicrucianism (p. 233).

- **Astrology:** Vallee, and later Hynek, became friends with Michel and Francoise Gauquelin, who were attempting to put astrology on a scientific basis. "Yesterday Hynek went back to see the Gauquelins to discuss astrology and destiny" (p. 341). Vallee claims he was responsible for his publisher Regnery accepting Gauquelin's book on astrology, *The Cosmic Clocks.* In the early days of CSICOP, there was a big stink when the skeptics challenged Gauquelin's "Mars Effect" data, which apparently was (in that one instance) correct. However, the correlations he claimed to find could not be replicated.

- **Mystical and psychic realms:** "In recent discussions with Hynek, I pointed out that the saucer question may well be part of a complex series of scientific realities, but it also plunges deep into mystical and psychic theories. I found him very receptive to this idea" (p. 88).

- **Alchemy,** elementals, homunculi, etc.

## The Flying Saucer Air Wars of 1952

Most people are totally unaware of what would surely be the most shocking event in U.S. military history, assuming that it actually occurred: the Flying Saucer Air Wars of 1952. At last this long-neglected history is finally revealed to the world in *Shoot them Down: The Flying Saucer Air Wars of 1952*, by Frank C. Feschino, Jr. The book, which boasts of a foreword (and an epilogue) by the "flying saucer physicist" Stanton T. Friedman, explains how the Air Wars were somehow connected to the famous Flatwoods Monster of Sept. 12, 1952, in which a twelve-foot-tall glowing hooded creature was reportedly seen floating through the woods of rural Braxton County, West Virginia. (Feschino explains that the creature's "hood" was just a misperception, it was actually wearing a space helmet.) Feschino claimed on the internet podcast *The Paracast* of July 10, 2007 that there were 18 ½ hours of UFO sightings that same day, from many different locations. Thirty objects were, he claimed, seen coming in over the Eastern Seaboard, and appeared to be following a damaged craft. Feschino claims that 180 military troops were sent to Braxton county to deal with all of the UFO crashes, and that a dead alien was found near Wheeling, West Virginia, all in that same night. The Flatwoods Monster, says Feschino, crawled out of one of those crashes, and it was dripping something like oil. Many pieces of metal were recovered from the crashes around Flatwoods; unfortunately, none of those pieces can now be located.

UFO crashes occurred that night not only in West Virginia, but in eleven states. However, Air Force F-94 Starfighter jets also went down near the UFOs. Stanton Friedman, in the same podcast, asserted that there has been a "hidden war" going on, resulting in plane crashes. As Friedman explained on the popular *Coast To Coast AM* radio show the night of December 7 2007, in the 1950s the US military had standing orders to shoot down unidentified craft if the craft didn't land when instructed— "and it appears that UFOs shot back." He added that there is no question that our planes were the aggressors. Friedman suggests that because of these losses the US military eventually gave up on their shootdown policy toward UFOs, and instead began simply observing them with their instruments. So it would appear that the Saucers' pilots, and not ours, are in fact the Top Gun on this planet.

## The Montauk Project: Molesting Little Boys to Warp Space-Time

Al Bielek, Preston Nichols, and Duncan Cameron have given a melodramatic account of alleged time-travel experiments that are said to have taken place at the old military base on Montauk, Long Island, New York in the 1950s and 1960s. They claim to have somehow slipped into the now closed-off area and down into a subterranean "chamber of horrors." They describe "cages" in which the unwilling subjects of the experiments—young adolescent boys—were kept. According to Bielek, Cameron, and Nichols, tens of thousands of boys were kidnapped off the street, put in cages, and repeatedly tortured, raped, and sodomized into total sexual submission. Why was this done? Apparently the mental shock of tortures, applied right at the moment of adolescent sexual climax, was discovered to be capable of opening up holes in the space-time continuum, allowing objects or even persons to pass through.

Later it was found possible to achieve the same effect with computers alone, by somehow "interfacing an IBM 360 with the Cray 1," which did not yet exist. When the two computers were supposedly connected, Cameron produced via his own mind-power an uncontrollable 25-foot monster that went storming about smashing buildings. All attempts to destroy the monster failed. Fortunately, it was sent into "hyper-space, to another time," By some accounts, the monster still runs amuck at night, when the moon is full and fog sits upon the moor, to menace unwary visitors.

Al Bielek and some other pioneers were once allegedly sent forward to the year 6737. There they saw a "golden horse" amid the apparent ruins of dead civilization. The intrepid voyagers were instructed to remain within a circle with a 20-foot radius, or else they could not return. Unfortunately, many of them were said to have been lost in time for eternity."

At at the UFO film festival at the 2012 *International UFO Congress* I watched *The Montauk Chronicles*, a documentary on these supposed sinister paranormal experiments carried out on innocent boys. The movie is quite well made. I met the writer and director, Christopher P. Garetano, who seems like a nice fellow but apparently lacks the gene for critical thinking. He cannot decide whether this huge, bloated, absurd tale is true or not.

## Apollo 18: The 'Moon Hoax' Stood on its Head

I'm sure the reader is familiar with claims of a "moon hoax:" that we never went to the moon, the Apollo program was a hoax, filmed on a movie lot, etc., etc. This has all been very ably refuted by many fine researchers - Phil Plait of Bad Astronomy, the Mythbusters, James Oberg, etc, and I won't go into the details of this nonsense. Suffice it to say that even Richard Hoagland, the promoter of the "Face on Mars" who believes every whacked-out space conspiracy you can imagine, agrees that the Apollo astronauts did indeed go to the moon.

But who says there's nothing new under the sun? Now a movie called *Apollo 18* suggests a new Moon Hoax, but the opposite of previous ones: not only did we go to the Moon, but we went more times than we admitted to - because we found aliens there.

The last Apollo mission was Apollo 17, launched on Dec. 7, 1972. Several more missions had been scheduled, but were canceled due to cost concerns. After Eugene Cernan and Harrison Schmidt blasted off the surface of the moon on Dec. 15, 1972, no human has set foot on the lunar surface, or even entered lunar orbit.

The premise of this movie - admittedly science fiction, but certain to be taken as fact by many - is that there was a "secret" moon mission after Apollo 17, which encountered aliens, and (as seen in the movie trailer), even brought back one to earth, presumably dead. I guess NASA would have to have already known that aliens were there, in order to run the Apollo 18 mission secretly. Maybe Apollo 17 is supposed to have found the alien evidence, but NASA didn't tell us, and went back in secret.

The movie trailer for Apollo 18 has to be one of the lamest things I've seen in a long while. It seems to show an astronaut blasting off from the Moon in a lunar lander, with what appears to be a dead humanoid aboard. The creature's head looks vaguely Mongolian, with a mustache, but it has female breasts. I'll refrain from making the obvious crude sexual comments, although others haven't ("space? boobs!!!").

The film is supposed to represent "found footage", as in the Blair Witch Project, Cloverfield, etc. I don't know if the film attempts to explain how NASA could possibly launch an additional Apollo mission in secret. How did nobody notice the massive Saturn V launch from Cape Canaveral? How did thousands of people worldwide perform their necessary support tasks (which they had done several times before, under great media scrutiny), and the news not leak out?

Well, it's entertainment even if it's lame entertainment.

## Jared Loughner, Conspiracy Fanatic

Details slowly emerged about Jared Loughner, the gunman responsible for the horrific shootings in Tucson on January 8, 2011, seriously injuring

Congresswoman Gabrielle Giffords and killing six others. While much political blame has been assigned, by all sides, the truth seems to be that Loughner did not belong to or sympathize with any organized or even halfway-organized political group. Unlike the Unabomber, Laughner's incoherent political screeds reveal no consistent political vision. But what they do reveal is a mind deeply steeped in conspiracy belief.

Loughner's Youtube videos (actually more like Power Point presentations accompanied by music) clearly reveal this conspiracy mindset. For example, he claims that "government is implying mind control and brainwash on the people by controlling grammar." His videos are filled with absurd and pointless syllogisms, such as

> If B.C.E. years are unable to start then A.D.E. years are unable to begin.
>
> B.C.E. years are unable to start.
>
> Thus, A.D.E. years are unable to begin.

More ominous was his short video about Mind Control: "I'm able to control every belief and religion by being the mind controller."

But it now appears that Loughner has been posting to at least one internet conspiracy site using the name "Erad3" (which would be an anagram of his first name if we substitute "J" for "3"). Erad3 posts some of the very same inane content that we find in the above YouTube videos, such as "infinite currency," and the pseudo-syllogistic style of both writers is clearly the same. It's virtually certain that Erad3 = Loughner.

On the UFO and other conspiracy-related site "**Above Top Secret**," **Erad3** began a long thread titled, "All aboard with the empty NASA Space Shuttles!" In it he argues not that astronauts never went to the moon, but never even went into space at all, using ridiculous "syllogisms" such as

> If the design of the NASA Space Shuttle keeps the black body temperature of -454 °F from the outside orbit then the NASA Space Shuttle is at a temperature for human life.
>
> The NASA Space Shuttle isn't at a temperature for human life.

Hence, the design of the NASA Space Shuttle doesn't keep the black body temperature of -454 °F from the outside orbit.

To their credit, the other conspiracy aficionados on that site argued with Erad3 fiercely. His conspiracy syllogisms obviously made no sense - this wasn't even a good conspiracy story, just the delusions of some madman. Erad3 also started a thread claiming that the Mars Rovers likewise were faked.

If NASA creates a mars rover that communicates from mars then the signal reaches from the distance of mars.

The signal doesn't reach from the distance of mars.

Nonetheless, NASA creates a mars rover that doesn't communicate from mars.

Space journalist and skeptic James Oberg brings up the possibility that this shooting may have been more than just blind anti-government anger. Congresswoman Gabrielle Giffords, who Laughner is accused of shooting, is married to NASA astronaut Mark Kelly. Oberg writes, "This raises the disturbing possibility that Giffords' husband, astronaut Mark Kelly, may not have been a coincidental feature of Laughner's delusions and murderous hatred. If he really had the belief that NASA was faking its space missions, then Kelly would have been one agent of that fakery -- and perhaps in his weird world, so would his wife."

Probably that was not the only reason for Laughner's mad anger, but likely contributed to it. Proximity - simple geography - could well have been the main one. It's easier to be upset with people who are nearby that one sees than those far away that one never sees. And for someone with obsessive beliefs about government mind control, unconstitutional government acts, and NASA space conspiracies, the "power couple" of Congresswoman Gabrielle Giffords, who represents his district, and astronaut Mark Kelly, a supposed participant in a huge NASA conspiracy, would represent everything that he opposed.

## Prevalence of UFO Conspiracy Belief

According to In a 1996 Gallup Poll, 71% of Americans agreed that "the U.S. government know(s) more about UFOs than they are telling us." In that same poll, 48% believed that UFOs are real. (A 1997 Time/CNN poll got similar numbers). Therefore, about 23% of the population believes that *UFOs are not real, but the government is covering them up, anyway*.

# *More people believe in UFO Conspiracies than believe in UFOs!*

# 7. GIVE ME DISCLOSURE, OR GIVE ME DEATH

Since the early days of the Flying Saucers, claims of government cover-ups and conspiracies have been widely made and widely believed, along with ringing cries that the alleged cover-up will soon be ended. Here are a few examples of predictions that the supposed cover-up would soon be ended:

- **"Large Grey Aliens Require Full Disclosure by 2015!"**

- **"COMING IN 2014 - ALIEN DISCLOSURE IMMINENT!"** (YouTube video, by Secureteam 10).

- **"Vatican UFO disclosure Soon"** (2009 Youtube video)

- "It is my analysis that the ending of the official government's UFO cover-up began August 7, 1996... we may expect further announcements related to UFOs and extraterrestrial life, after the November 5 election." UFOlogist Dr. Richard Boylan

- "Before the year is out, the Government perhaps the President— is expected to make what are described as 'unsettling disclosures' about UFOs" - U.S. *News & World Report,* April 18, 1977.

- "Aliens... will begin transmitting their secrets to us no later than August, 1977" - Jeane Dixon, 1976.

- "We predict that by 1975 the government will release definite proof that extraterrestrials are watching us." - Ralph and Judy Blum, in *Beyond Earth: Man's Contact with UFOs* (1974).

- "The time is getting near when the U.S. Air Force will have to end its longstanding tactic of concealment." - Syndicated columnist Roscoe Drummond, 1974.

- "FLYING SAUCERS—THE REAL STORY: U.S. BUILT FIRST ONE IN 1942. Jet-propelled disks can outfly other planes ... By choosing which [jet] nozzles to turn on or off and the angle of tilt, the pilot could make the saucer rise or descend vertically, hover, or fly straight ahead, or make sharp turns... a big advance in the science of flying... No official announcements are being made yet, but about the only big secret left is "who makes them." Evidence points to Navy experiments... " - News "scoop" in *U.S. News & World Report,* April 7, 1950.

If the social phenomenon of UFOs tells us anything, it is that the future of the movement turned out differently than its proponents expected. For at least twenty years after Kenneth Arnold's sighting, believers expected that sometime soon, any day now really, a UFO would land openly—or would crash and be recovered—or otherwise be indisputably revealed. At the very least, believers hoped, the Air Force would end its alleged cover-up of the data it held about UFOs and disclose that information to the public. By the 1970s, this expectation changed. With mass sightings having gone on for over twenty years with no tangible result, UFOlogists' hopes transferred to UFO abductions providing the desperately sought Holy Grail of proof. When abductions had gone on for over thirty years without producing anything tangible, excitement shifted to claims of crashed saucers. The idea of a major "disclosure" coming soon has long been a major hope and expectation in UFOlogy, paralleling Christian fundamentalists' expectation of the Second Coming.

In the March 1991 issue of *Fate* magazine, UFOlogist Jerome Clark reviewed two new books on Roswell and excitedly predicted: "Major media—not just the usual tabloid papers—will pick up the story and recount their own investigations, which will confirm the UFOlogists' findings." Of course, this never happened. More than three decades have passed since the publication of *The Roswell Incident*, and the case has sustained heavy blows by the findings about Project Mogul. Subsequent alleged saucer crashes never achieved anything near the level of belief or publicity that Roswell did (at least something *did* crash near Roswell, even if it wasn't a UFO). So it's likely that UFOlogy is ready for the "next big thing." What that will be is difficult to say. Skeptical researcher Martin Kottmeyer has famously described the UFO movement as "an evolving system of paranoia," and as such it's difficult to predict where its paranoia will evolve next. Whatever it may be, we can expect it to offer an element of personal relationship or involvement (like contactees and abductees), to sound exciting and at least a little dangerous, and above all to promise such stunning evidence as to blow the alleged cover-up sky-high. It will have to excite and to entertain simultaneously—a tall order, but one that UFOlogy has always been able to fill.

## The Disclosure Project

Dr. Steven Greer, M. D. is the leader of CSETI, the Center for the Study of Extraterrestrial Intelligence, which has nothing to do with SETI. It's a wild-and-woolly UFO group that claims to be able to lure spacecraft down for close encounters by shining lights at them. If that doesn't work, they try "coherent thought sequencing", an exercise in group telepathy, which apparently gets 'em every time - see *www.cseti.org* for the whole story. Greer claims to have had "hundreds" of close encounters: in November of 1998 he claims to have seen a dozen spacecraft, clustered up in the stars. As for seeing actual ETs themselves, that's less common: he's only seen them a few dozen times. When Greer moved from North Carolina to Charlottesville, Virginia, alien activity followed closely in his wake. According to the *Charlottesville Weekly* (Feb. 2, 1999), "Greer had just bought a house in Albemarle County when the circles appeared here." Greer noted that the crop circles were in the shape of the CSETI logo. "It was a welcome mat they - the extraterrestrials - put out for the director of CSETI," he modestly explained.

Greer is a big promoter of the gospel of reverse-engineered alien technology, first proclaimed in *The Day After Roswell* by the late Philip Corso. "We possess and have reverse-engineered functional extraterrestrial devices that operate with physics not being taught at UVA. I have seen them." But don't assume that Greer is naively credulous: he thinks that at least 90% of the claims of UFO abductions are "absolute rubbish or hoaxes." As for the rest, he thinks it's due to a military program using craft that look like UFOs, in which people actually are abducted (which incidentally was the plot of an episode of The X-Files).

One group working to uncover the supposed Grand Conspiracy is Greer's Disclosure Project. They claim to have assembled over 400 military and government witnesses to UFO events and projects who are willing to give testimony about them. "The weight of this first-hand testimony, along with supporting government documentation and other evidence, will establish without any doubt the reality of these phenomena" according to Greer. This evidence was presented to the media in a much-hyped press conference at the National Press Club in Washington, D.C., on May 9, 2001, where "more than 20 military, government and corporate witnesses to unambiguous UFO and extraterrestrial events stated their testimony before millions." The event was streamed live on the internet, resulting in the largest audience by far for any web cast from the National Press Club, overloading the servers.

Many elements of the news media gave largely uncritical accounts of the event, leading the public to suspect that the claims might well have some validity. In a story dated May 11, the *ABC News* website headlined "Former Government Employees Say It's Time to Reveal Evidence," adding "They're out there — and the government knows." *Fox News* reported on May 10, "A group of 21 former military and government officials told a packed house at the National Press Club stories worthy of a whole episode of the *The X Files* Wednesday. These men in suits spoke of high-speed saucers, crashed ships, alien bodies and conspiracies of silence." While the BBC coverage suggested at least some degree of humor and doubt, *The Independent* reported, "Yesterday 20 witnesses from the American military, intelligence services, and scientific establishments gave their testimony to start a campaign which they hope will force the government to investigate the UFO phenomenon." Even Russia's *Pravda* jumped into the fray, its headline of May 12 proclaiming "American Scientists Demand to Publish the Information about UFO Accumulated During 50 Years." The only really skeptical account outside the skeptical movement came from *Wired* magazine: "Ooo-WEE-ooo Fans Come to D.C."

However, after the cover-up allegations of Greer and his colleagues were repeated widely on all the news outlets for at least a few news cycles, they simply disappeared, failing to convince even the most sensation-hungry reporter that there was pay dirt under the dust and chaff. Not one of the "disclosure witnesses" could produce a single shred of evidence beyond their own unsupported words, and many of them carry such baggage that believing what they say becomes a Herculean task

Greer then took his show on the road, giving "Campaign for Disclosure" presentations in many North American cities including San Francisco, Miami, Los Angeles, Toronto, New York, and Vancouver. In each city, the show began with "an exclusive showing of the 2-hour Disclosure film of 50 government and military witnesses to UFO and Extraterrestrial events and projects," followed by a presentation from Greer.

What the media uniformly failed to report were the many absurd statements that were a dead giveaway to anyone having even the least education that the entire event was an exercise in surrealism, and not "disclosure." Greer's claims of secret technology involving "zero-point energy," "anti-gravity," and even "superluminal" devices should have been red flags to any knowledgeable reporter. My favorite howler was Greer's matter-of-fact statement that "superluminal" (i.e., faster-than-light) alien craft had been reverse-engineered by American scientists, and were now being manufactured by Lockheed-Martin. If Greer is correct, not only is there a secret government covertly calling the shots, but also a secret realm of earthly technology that has ready access to undreamed-of capabilities, based upon a secret physics that knows all about "superluminal" travel and other miracles, but whose priesthood has managed to keep the astonishing news completely hidden from their students, mentors, and other professional colleagues. (The first two I find extremely difficult to believe, the last one impossible.) Compounding the Project's credibility problem is the great emphasis on 'free energy' claims, that "the internal combustion engines, the fossil fuel plants, gas, oil, coal, and ionizing nuclear technology" are all obsolete, a fact that the government is working very hard to cover up to protect the vested interests of those industries. Greer is featured prominently in the bizarre 2011 conspiracy-oriented documentary movie *Thrive*, promoting 'free energy' claims. Greer's own 2013 movie *Sirius* is filled with similar claims.

Sergeant Cliff Stone testified about having participated in the recovery of a crashed extraterrestrial spacecraft in Indian Town Gap, Pennsylvania in 1969, complete with alien casualties. (Somehow the people in this town have been unable to capitalize on their very own UFO crash as the folks have in Roswell.) Stone earlier claimed to have seen crashed saucers in the basement of the Pentagon. He also claims that while he was serving in Vietnam, he was sent to recover a downed B-52 that looked as if it had been simply "plucked from the sky and set down in the jungle" by a UFO.

Daniel Sheehan is well known as a lawyer for progressive causes. He was involved in the Pentagon Papers case, the lawsuit over Karen Silkwood's death, and many other high-profile cases. He headed up a group called the *Christic Institute*, which in the 1980s filed wide-ranging lawsuits claiming that the CIA had set up a giant 'secret government' based on drug smuggling, assassination, etc. Apparently the CIA was also involved in the assassination of JFK. This lawsuit made big news as it dragged through the courts for years, deposing large numbers of government officials, until the courts finally dismissed it as having no merit. Even prior to Sheehan's recent "recovered memory" of having been shown the proof of captured UFOs, his claims of CIA conspiracies had already earned their niche in Jonathan Vankin's book *The Sixty Greatest Conspiracies of all Time*. But Sheehan didn't say a word at that time about his fantastic UFO discoveries: he claims to have seen the "proof" of secret crashed saucers during the closing months of the Ford administration in 1976, but he said absolutely nothing about it until the year 2001, in spite of spending much of the 1980s and 90s in front of news cameras, beating the drum about bizarre government "conspiracies." Were it not for this sudden and fortunate recollection of his astonishing experience, Sheehan might yet still be completely out of the news and the public eye.

Dr. Carol Rosin is a long-time leftist activist who claims to have been a spokesperson for the late Dr. Werner von Braun. She, too, seems to have very recently discovered the significance of UFOs. She had been a long-time campaigner against all space-based weapons, whether offensive or defensive. Her role in the *Disclosure Project* was "to testify that the early space pioneer warned of the dangers of space-based weapons and that covert programs were planning to eventually justify them by manufacturing a threat from outer space which does not exist." She did not actually make any UFO claims of her own.

Former U.S. airman Larry Warren is well known in UFO circles as a supposed witness to alleged 1980 UFO encounters at Rendlesham in the UK. He claims to have seen Air Force officials meet with extraterrestrials, who floated in the air, surrounded by a luminous protective bubble. At the *Disclosure* press conference, Warren told of allegedly being debriefed after his encounters by being shown a secret government film. Not only did the film reveal the existence of extraterrestrials, it even showed Apollo astronauts on the moon pointing out artificial alien structures.

Astronaut Gordon Cooper told how he had seen a film taken of a UFO landing at Edwards Air Force Base, California, in the 1950s. The film was allegedly sent to Washington, but subsequently "disappeared." Greer comments that "his testimony cannot be contradicted because of the credibility that this great man has." Perhaps, but it also cannot be substantiated, and Cooper's "credibility" is seriously compromised by his history of making extraordinary claims that don't check out (**James Oberg** has done an in-depth analysis of **Cooper**'s **UFO** claims).

Greer claimed Apollo astronaut Edgar Mitchell as one of his "disclosure" sources. Goodness knows that Mitchell has certainly endorsed enough *other* paranormal absurdities so that his participation here would hardly surprise us. However, Mitchell has stated unequivocally, "I cooperated with Steve Greer some years ago, but he began to overreach his data continuously, necessitating a withdrawal by myself, and, I believe, several others. I have requested to be removed from any web site, announcements, etc., but see that has not taken place.... [neither] I, nor any crew I was on (I was on three Apollo crews), received any briefing before or after flights on UFO events, saw anything in space suggesting UFOs or structures on the moon, etc." Nonetheless, Mitchell's name still remained on the "Disclosure Project" website as a supporting witness.

As you move down Greer's list of 400 allegedly supporting witnesses, it does not get any more convincing. Gene Mallove was well-known as a cold-fusionist. When Mallove was murdered in his home on May 14, 2004, apparently during a robbery, many insisted that this was a 'hit' ordered by the Powers That Be to silence him. Paul LaViolette calls himself the 'new Einstein.' Thomas Bearden has been writing bizarre papers about "noncausal phenomena" and other mumbo-jumbo for decades. Glen Dennis is a former mortician whose claims to have seen alien bodies at Roswell has been shown by skeptic Philip J. Klass to be "riddled with flaws and inconsistencies."

Other gaping holes in the Disclosure fabric were soon spotted. British researcher James Easton noted that Greer's "executive summary" contains a number of demonstrably false claims: A woman who claimed witnessing UFO activity at the RAF base in Bentwaters, UK in 1980 had in fact left the base well before that incident, and Greer also repeats the false claim that astronaut Gordon Cooper had filmed a UFO in space, but the film later 'disappeared.' Oberg noted that Greer's witness Donna Hare, who claims having seen NASA airbrush out an image of a UFO in a photo before releasing it, also claims to have seen trees and their shadows on the ground in these pictures taken from orbit – a physical impossibility, given their low resolution.

Lara Johnstone of Berkeley, California was so moved by Greer's "Disclosure" that on July 28, 2001 she began a hunger strike to demand full 'disclosure,' allowing herself to drink only water, vitamins, and juice. "I am asking President Bush to fulfill his campaign promise, and make public information on U.S. government involvement with Extraterrestrial craft, technologies, and civilizations by supporting the Disclosure witnesses who are ready to testify before the U.S. Congress about their extensive knowledge of the Extraterrestrial presence," she wrote. Despite the publicity Johnstone received in the UFO press and on Art Bell, the president ignored her repeated urgings for a response. Finally on Sept. 9, day 44 of her hunger strike, Johnstone called it off by eating a burger – "a small vegetable one," she hastily added.

Greer sank so low as to attempt to exploit the tragedies of September 11 to advance his sagging cause. "We Need Your Help to DISCLOSE the SECRETS that could have PREVENTED the SEPT 11th TRAGEDY. Hear how Classified Projects are withholding Technologies that could Replace our need for Oil," proclaimed the Disclosure website. "The good news is that the Disclosure Project can prove that we have a replacement for oil, coal and conventional energy. In a decade we could reach energy independence: bin Laden *et al* may keep their oil, for we will not need it." The bad news is that these supposed 'energy replacements' are no more real than Greer's wild saucer claims.

Indeed, so absurd are many of the claims of Greer and friends that a substantial backlash against them has been underway within UFOlogy. Pro-UFO researcher Robert Swiatek has noted Greer's "often radical and unsupported pronouncements, as well as the considerable problems that they have created for serious, scientifically-inclined UFO and abduction researchers." The non-conspiracy UFOlogist John Alexander, in his book *UFOs Myths, Conspiracies, Realities* says that by 2001 a number of UFO activists including himself working behind-the-scenes were building support in Washington for a new round of Congressional hearings on UFOs. (An earlier round of hearings in the 1960s generated headlines, but resolved nothing.) They believed their goal would soon be reached. Then along came Greer with his absurd, highly publicized claims, making sensational cover-up accusations against the government, and suddenly all possibility of government support for an investigation into UFOs was lost. Hence Alexander's observation that "the UFO movement is its own worst enemy."

Many UFOlogists are also troubled by Greer's contention that the aliens are friendly toward humans, or at least not hostile. It is widely accepted by many "abduction" researchers that the motivations behind UFO abductions are positively sinister. For example, David Jacobs writes in *The Threat* that

> the abduction phenomenon is far more ominous than I had thought. Optimism is not the appropriate response to the evidence, all of which strongly suggests that the alien agenda is primarily beneficial for them and not for us. I know why the aliens are here -- and what the human consequences will be if their mission is successful.

From this perspective, Greer's 'peaceful alien' scenario is not only misleading, but fraught with peril. Michael Brownlee writes, "it appears that one of the hidden objectives of Greer's CSETI disclosure effort may be to deflect investigation away from the extraterrestrials themselves. They remain almost completely absent from his disclosure documents. Rather than entertain questions or evidence about the ETs, Greer is quick to deflect such discussion. He consistently responds to any concern about ET motives or intentions with his standard scathing reply: "There is no evidence that any of them are hostile. So therefore there is no objective threat." Thus while some UFOlogists worry that the obviously low credibility of the Disclosure Project will wrongfully taint the public perception of "serious" UFOlogy, others are concerned that Greer's sanguine estimation of alien intentions leaves us unprepared for the coming alien holocaust, once their crossbreeding program has been completed.

*The author's photo of Dr. Steven Greer getting a big hug from one of his fans, Leda Beluche (GirlieVegan.com)*

Finally in 2009, the world was allowed to see the **"First Ever Photograph of an Extraterrestrial**," courtesy of **CSETI**! Of course, there have been a number of *alleged* photos of Extraterrestrials in the past, but apparently all of those were fake and CSETI has snapped the first *authentic* one. The photo was reportedly taken on Nov. 17, 2009 in Joshua Tree National Park, in the California desert east of Los Angeles, not far from where the late, great George Adamski used to go into the desert to meet up with his Venusian pals. (Aliens seem to like deserts, except when their saucers crash there.)

The "CSETI staff and a group of 40 national and international students spent six nights deep in the park, learning and practicing the contact protocols... a series of remarkable - and historic - events transpired, including the visit of an extraterrestrial being within a few feet of the contact team, photographed by a senior team member." According to CSETI, "This photograph offers extraordinary evidence of interstellar, transdimensional technology, and the efficacy of CSETI contact protocols." Greer writes "You will note that the ET is suspended in a cone of light which is originating from a small orb to the left of the bush... The ET appears to be a male, wearing a type of vision augmenting goggles, with a very large head with an indented area demarked by ridges in the forehead. The hairline, ears, eyes, mouth and chin are clearly visible. Both arms can be discerned, as well as a torso and both legs, with boots on the feet. He is hovering a foot or two above the chairs that make up our contact circle, and is just east-southeast of the circle. His size is estimated at 3-5 feet in height. Note that he is leaning forward, with his torso and head twisted to look directly at the camera. His right leg is bent behind him." Unfortunately, "Neither the orb nor the ET being was seen with the naked eye," but Greer goes on to explain how cameras can sometimes record objects that the eye cannot. The website suggests, "These photos look much clearer when viewed on a High Definition computer screen or a High Definition TV screen." To me it just looks like a confusing blur, no matter how many times I enlarge it. Similarly, I can't make out all those alien artifacts people see in photos from the Moon and Mars. I think I need a new monitor.

On October 2-4, 2010 Greer's organization CSETI held a seminar in Rio Rico, Arizona, where for just $495 (room fees not included) he promised to teach conference registrants his techniques of how to signal ETs for contact. (The fee for a week-long intensive CSETI program has now risen to $2,500) According to one participant, these "ET Ambassador Training" sessions were supposed to take place from 8:00 to 12:00 nightly, but the ETs failed to show. One might think that such embarrassing failures would make it hard to get people to sign up for future seminars, but Greer confidently scheduled another seminar in Joshua Tree, California for the following month. He knows that there's one born every minute. Participants are required to sign a non-disclosure agreement, lest Greer's secret methods become known to the public. I spoke with a man who participated in one of these desert sessions in 2012. He explained how the participants were unable to see the extraterrestrial ships and beings at the beginning of the session, until Greer and his associates taught them how to do so. They gradually acquired this skill, and were quite proficient at it by the end of the session.

At the 2012 International UFO Congress, Greer spoke on "Contact: Countdown to Transformation" to his largely-credulous and adoring crowd. "Disclosure has already happened," said Greer unexpectedly, borrowing a line from John Alexander, who Greer normally disagrees with about everything. Greer didn't explain that remark. Greer showed a photo of Bijou, an ambassador from the Andromeda Galaxy whose acquaintance was made in one of CSETI's desert Skywatches at Joshua Tree, CA. Bijou has now become like a sort of pet alien for Greer, often playing peek-a-boo when it's least expected. It takes a great deal of imagination to see Bijou in the original photo, even after they have helpfully cut away his outline (right) - more imagination than I have, I'm afraid. Greer also has equally-fuzzy photos of blips that he says are alien spaceships, many of them only partially materialized in our dimension. Every time they go out, says Greer, they spot UFOs. Every time. This confirms my suspicion that no matter what light they might see in the sky, they say it's a UFO.

Greer claims that electro-gravitic devices were built by as long ago as the 1950s. Alien technology, he says, offers us free energy so that we won't need oil or coal any more. However, this is being opposed by the greedy oil companies, and the military-industrial complex. Greer said that he was offered a bribe of $2 BILLION by the conspirators to stop investigating UFOs, which he bravely and selflessly declined. (Me, I'd sell out for a mere two million, but then I don't make as much money as doctors do). He said the only reason the conspirators had not yet killed him is because he is too well-known. He accused the conspirators of "murder" in the cancer death of his friend Sheri Adamiak, and in the non-fatal cancers of other researchers, including his own. Apparently they have some sort of 'cancer ray' that they use against UFOlogists who get too close to the truth. There was this Rambo guy that I saw with Greer when he first arrived at the conference. Rambo was dressed in black combat fatigues, and appeared to be wearing a bulletproof vest. I thought this was odd, but I didn't pay much attention to him. Arizona is a state where people carry around guns the way people in other states carry umbrellas, although there were "no weapons" signs posted all around the conference facility. I heard later that Greer has taken to traveling with bodyguards, undoubtedly for the theatrical effect. Rambo, however, was not to be seen on Saturday when Greer spoke, and especially when he went into the hallway to sign autographs, which is when Greer would need protection the most. For whatever reason, Rambo wasn't there, leaving poor Steven Greer defenseless. I am happy to report that Greer survived his stint at the UFO Congress.

At **Steven Greer'**s lecture at the **2013 *Contact in the Desert*** conference at the Joshua Tree conference center, California, he had three security men, wearing yellow hats, positioning themselves in from of the exits, and they remained there for the duration of the lecture, for unknown reasons. We can see them in a YouTube video of part of his lecture. I have never seen security guards do this at any other conference. David Wilcock, another conspiracy-oriented lecturer at that same conference, reportedly used his security guards to prevent anyone from leaving during his lecture! Wilcock also reportedly brought a bomb-sniffing dog to check out possible bomb threats. This exercise in Security Theater seems to be quite effective in impressing the credulous.

In July of 2012, Greer announced that "The Disclosure Project and CSETI has teamed up with Emmy award winning filmmaker Amardeep Kaleka to make an historic new documentary on Disclosure, Contact and the suppression of New Energy." However, "No major studio or media group will touch this story: It is simply too explosive and world- changing for large corporate interests to embrace." (Why am I not surprised?) So he is collecting nickels and dimes (and presumably plenty of dollars, too) via crowd-sourcing from ordinary folks just like us. And he was already very close to meeting his goal. The film was described as follows:

> The Earth has been visited by advanced Inter-Stellar Civilizations that can travel through other dimensions faster than the speed of light. What we have learned from them about energy propulsion can bring us to a new era, but those in power have suppressed this information in order to keep us at their mercy.

By July 28 Greer was explaining on his Blog that the original funding goal for his movie had been met, but even more money would be needed:

> There is a chance that we may be able to include in the film "Sirius" the scientific testing of a possible Extraterrestrial Biological Entity (EBE) that has been recovered and is deceased. This EBE is in the possession of a cooperative institute desiring further scientific evaluation of the possible ET. We cannot reveal at this time the location of this being or the name of the person or persons who possess it.

> Dr. Jan Bravo - who is a STAR Board member and a fellow Emergency Physician - and I have actually visited the group that possesses this EBE and have personally and professionally examined the being. It is indeed an actual deceased body, and most certainly is not plastic or man-made. It has a head, 2 arms and 2 legs and is humanoid. We have seen and examined X-Rays of the being. Its anatomy however is not homo sapien (modern human) or any known hominid (predecessors to humans).

As you can imagine, the security and scientific issues surrounding the further testing of this potentially explosive and world- changing evidence are mind-boggling. However, we feel we simply must proceed expeditiously but cautiously. The cost of doing proper MRI testing, full and dispositive forensic-level DNA testing and carbon dating with other isotope testing are considerable and certainly not currently funded. We must rule out other hominids, bizarre genetic defects and so forth. But it is most certainly an actual biological specimen – and it may be – well, what it looks like.

Greer's very own dead ET – in his movie!

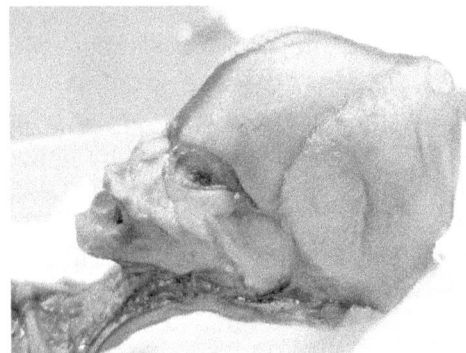

*The Atacama humanoid - Greer's supposed alien.*

Greer's movie *Sirius* was released in April, 2013, promising Free Energy and a deal alien. As the film opens, we see Greer going into a college auditorium in Santa Monica, the audience being checked with metal detectors for weapons. "Most people don't know what a Dead Man Trigger is." Very few people need one. But Greer has one - if the Conspiracy rubs him out, lots of sensitive documents get sent out to influential people.

The movie explains how certain crank free-energy "inventors," plus ET technology, offers us unlimited Free Energy, but a conspiracy by those Greer calls the "Petro-fascists" keeps us using coal, oil, and nuclear power. Part of the Conspiracy is to keep us distracted by other things. Even Honey Boo-Boo is depicted as part of the Conspiracy to keep us distracted from ET truths.

"We have acquired an EBE! (Extraterrestrial Biological Entity)", boasts one of Greer's CSETI colleagues, and the analysis of this little guy - just six inches long - is the principal "news hook" for the film. We are told that this little body was dug up in the Atacamba Desert in Chile, and ended up in Barcelona, Spain. 3D scans reveal its internal organs, apparently very human-like. Now here is the biggest "bombshell" that the film has: DNA from this creature has been analyzed by Dr. Garry P. Nolan of the Stanford University Medical School. He concluded that

> The specimen was concluded by the medical specialist to be a human child with an apparently severe form of dwarfism and other anomalies... Reconstruction of the mitochondrial DNA sequence and analysis shows an allele frequency consistent with a B2 haplotype group found on the west coast of South America, supporting the claimed origination of the specimen from the Atacama Desert region of Chile.

In other words, the "bombshell" revelation in Greer's film is that DNA analysis shows that the much-hyped 'humanoid creature' is actually a deformed human specimen suffering from severe dwarfism and other deformities. And people contributed via crowd-sourcing to reveal that!

## Dan Burisch and J-Rod the Alien

There has been much controversy over the claims of one Dr. Dan Burisch, who is supposed to be a PhD microbiologist who has worked on designer diseases for a rogue black-ops government program at Area 51 in the Nevada desert – and who also studied the extraterrestrial beings that are in residence there. Burisch claims to have worked alongside an alien gray named J-Rod 52, whose race hails from Zeta Reticuli. However, they are somehow supposed to be "time travelers" from our future. Burisch warned of a sinister conspiracy between E.T.s and the military to develop deadly diseases to wipe out humans that the conspirators want to be rid of. He claimed to have been in "lock-down" at Area 51, and requested Congress grant him immunity in return for his testimony about this nefarious program. Some UFOlogists cautiously supported Burisch's tales, others supported the stories even more strongly while accusing the first group of distorting them, and yet a third faction angrily charged that the Burisch supporters were disinformation agents charged with "spinning" wild tales to sow confusion within UFOlogy. (UFOlogists seldom believe that other UFOlogists with whom they disagree are simply mistaken – they are usually either pathological liars, or else government agents.)

Research by George Knapp and others published in the August, 2004 issue of *UFO Magazine* has pretty well sent Burisch's credibility crashing down faster than a UFO daring to fly over New Mexico. Burisch, whose real name is Dan Crain, claims to have earned a PhD at the State University of New York in 1989, but the school has no record of that. During the time he claims to have been earning his PhD in New York, he was in fact working as a parole officer in Las Vegas. But UFO journalist and researcher Linda Moulton Howe, one of the biggest supporters of Burisch's claims, says that she still believes his wild story about working with aliens because of his "sincerity."

Burisch claims that J-Rod 52 was sick, so he carefully took tissue samples. Hoping to use J-Rod 52 to communicate with his home planet, the rogue block-ops conspirators had him taken to Abydos, Egypt, where there supposedly exists a naturally-occuring StarGate. However, instead of helping E.T. to phone home, Burisch pushed him through the Stargate, back to the extraterrestrial family and home he so desperately missed.

## Will Hack for Extraterrestrials

Gary McKinnon of London was arrested in 2005 for computer crimes, after a three-year investigation. He has been described as "the world's biggest computer hacker." McKinnon reportedly caused $1 million in damages by hacking into at least 92 secure networks at NASA and the Pentagon, over a period of twelve months. American officials first thought McKinnon's hacking to be the work of Al Qaeda operatives. He is said to have deleted "critical" files from U.S. computers, stolen codes, and deleted 1,300 user accounts, among other acts of computer vandalism and larceny.

What motivated McKinnon's single-minded frenzy? He was convinced that the U.S. government was engaged in a massive cover-up of UFOs, and set out to prove it. Andrew Edwards, an old school chum of McKinnon's, explained "Gary told me all he was doing was looking for proof of a cover-up over UFOs." McKinnon, who claims to have been motivated by what he heard in the Disclosure Project press conference of May 9, 2001, found a whole lot of secret files. He told *The Guardian* (July 11, 2005) that he found evidence in the U.S. Space Command network of an "extra terrestrial mission." MacKinnon claims, "I found a list of officers' names under the heading 'Non-Terrestrial Officers'... What I think it means is not earth-based. I found a list of 'fleet-to-fleet transfers' and a list of ship names. I looked them up. They weren't US Navy ships. What I saw made me believe they have some kind of spaceship, off-planet." However, he said he couldn't remember much about his amazing discovery, because he was "smoking a lot of dope at the time". MacKinnon never did find proof of any UFO cover-up, and for several years he faced the possibility of a long prison sentence.

The appeals on his case kept McKinnon in legal limbo for a full ten years. Finally in 2012 the U.S. withdrew its extradition request to try him for computer crimes, in part because doctors had diagnosed him with an autism spectrum disorder compounded with clinical depression. Gary McKinnon is again a free man.

## Alien Invasions!

Nick Pope is making an entire career out of having worked for the UK's Ministry of Defense UFO investigations from 1991 to 1994. (He claims to have been in charge of UFO investigations, but his actual title was "junior desk officer," and he wasn't in charge of anything.) One would think that such a brief stint would not count for much, but Pope is not one to let an opportunity pass. He gets invited to all the major UFO conferences, and frequently is a guest on the *Coast to Coast AM* radio show, to spin his tales.

In March 2012, **Pope** claimed that a **UFO photo** that was perhaps the best and clearest ever taken '**mysteriously vanished**' from the MOD offices, where it had been hanging on the wall. In June, 2012 Pope got a lot of media attention when he warned that **mass sightings of UFOs** were likely during the upcoming **London Olympics**. When some videos showing the Goodyear blimp over the Olympic stadium were interpreted by some people as a UFO (see Chapter 3), some said 'Nick Pope predicted this.' (Of course, it was only people seeing the video on TV who called the blimp a UFO; people in the stadium recognized it as a blimp, and did not report seeing a UFO.)

Pope also warned that "The government must - and has planned - for the worst-case scenario: alien attack and alien invasion. Space shuttles, lasers and directed-energy weapons are all committed via the Alien Invasion War Plan to defence against any alien ships in orbit," he said, apparently unaware that America's Space Shuttle orbiters had already been dispersed to various museums, and the rest of the system scrapped.

When the collective response among "serious" UFOlogists was astonishment, Pope began to slowly backtrack, in a very clumsy manner. He told Richard Dolan (who says quite a few wild things himself, like about aliens on Mars),

> My comments concerning Alien Invasion etc, arose because I was commissioned to do some tie-in PR for the launch of the alien invasion themed Sony Playstation game, "Resistance: Burning Skies", out exclusively on the Playstation Vita. I came up with the idea of an alien invasion war plan. There are 2 versions out there, I can email you copies, a serious version and a more pop culture version that got picked up by the *Daily Mail*. The real issue here, and it applies to lots of subjects, is the increasingly blurred lines between real news reporting and marketing.

In an entry titled The End of Nick Pope, blogger "nickpost" on ufodaily.net revealed his own conspiracy theory:

> Nick Pope has completely discredited himself in the field of UFOlogy. Ironically he speaks of "the real issue of blurring the lines between news and marketing" whilst perpetuating just that. *It also reconfirmed my assumption that he's not ex Ministry of Defense.* The issue of Disinformation and Counter Intelligence Programs, (COINTELPRO ), designed to *deliberately spread false Information in order to keep the public away from the truth* and muddy the waters by intentionally spreading false stories to de-legitimize the subject. In other words if some of the info is true and some of it is ridiculous, it's all perceived as ridiculous. [emphasis added]

Blogger Simon Sharman writes, "In Pope's own words he 'came up with the idea of an alien invasion war plan' only because he had been commissioned to do some PR for Sony. Secondly, his war plan was carefully thought out because he has no knowledge that any such plans exist." Or in other words, he made up the claim about a government plan for alien invasion, for commercial gain.

UFOlogist Joe McGonagle wrote, "I struggle to understand why anyone thought Nick Pope had any credibility to start with":

# Give Me Disclosure, or Give Me Death!

The same man who claims to have investigated Alien abductions, crop circles, and animal mutilations for the British Government when he was "in charge of" "the British Government's UFO Project", when according to the head of his department in 1997; "Turning specifically to your comments concerning Mr Pope, I should point out that he was a junior desk officer in the Secretariat (Air Staff) 2a section from 1991-1994 and was not in charge of, or the head of any part of Secretariat (Air Staff) 2. Mr Pope was an executive officer and shared the support of one administrative officer" according to one of his successors, Linda Unwin, "The first point to make is that there is no 'UFO Project'. Handling of UFO sightings is a very small element of our work."

McGonagle adds that Pope is "the man who still pushes the "Cosford incident" as unexplained, when in fact there is an obvious explanation for the majority of the reports on 31st March 1993...[the re-entry of the rocket booster that launched the Cosmos 2238 satellite] The same man who discussed an obvious image of a gull as "If I was still there [on the UFO desk] I'd be looking at this very closely. The object looks structured, symmetrical and metallic"... The same man who continues to portray the radiation readings as hard evidence of something unusual at Rendlesham forest when in fact the readings are meaningless."

Some have described Pope as a 'very conservative' UFOlogist, but the facts don't bear this out. As skeptic Ian Ridpath noted, Pope "has been banging on about potential alien invasion for years." In a news interview with *The Daily Mail* (Nov. 10, 2006), **Pope** warned, **'Aliens could attack at any time'**:

a former MoD chief warns that the country could be attacked by extraterrestrials at any time... During his time as head of the Ministry of Defence UFO project, Nick Pope was persuaded into believing that other lifeforms may visit Earth and, more specifically, Britain.

His concern is that "highly credible" sightings are simply dismissed. And he complains that the project he once ran is now "virtually closed" down, leaving the country "wide open" to aliens.

There has been no suggestion that Pope was 'promoting a video game' at that time. And if you do a search on **Nick Pope alien invasions**, you'll find several more examples.

## Exopolitics – the "Political Implications of the Extraterrestrial Presence"

Another popular exercise in contemporary UFO fantasy involves what is called *exopolitics*, "political implications of the extraterrestrial presence." It is the brainchild of Michael Salla, who, with a doctorate in government, traveled worldwide to participate in conferences and retreats, campaigning for peaceful relations between humans and extraterrestrials and for an end to the alleged UFO cover-up. Since there is no actual evidence of any alleged "extraterrestrial presence," exopolitics has much in common with medieval disputes concerning angels and pinheads. Nonetheless, exopolitics has become a significant player on the UFO scene, and Salla's Web site, *exopolitics.org*, receives several million visitors yearly. Exopolitics claims that "hidden agreements concerning extraterrestrial life have been secretly entered into by a range of government-authorized agencies, departments, and corporations. In some cases, these pacts involve representatives of advanced extraterrestrial civilizations whose existence has not been disclosed to the general public." They insist that such agreements should be made openly.

I've written quite a lot about Exopolitics in my *Psychic Vibrations* column, and I almost feel bad for doing so. The claims made in Exopolitics are so transparently absurd that refuting them is like shooting fish in a barrel, and poking fun at exopoliticians sometimes feels unsettlingly close to the 18th century practice of visiting Bedlam on a Sunday afternoon to poke sticks at the inhabitants. Every time you turn around, exopolitics has some hilarious new absurdity, such as Alfred Lambremont Webre's claim that people are being teleported back and forth to a secret American base on Mars. Webre is described as a "lawyer, cosmologist, and futurist, was the founder of Exopolitics and is one of the leading commentators of the Time-Space Age." But in reality Salla can probably claim the dubious honor as the "founder" of Exopolitics. Webre has the website exopolitics.com, while Salla has exopolitics.org.

Salla, who was educated in Australia and has degrees in government and in philosophy, was a Researcher in Residence in the Center for Global

Peace, American University, in Washington, D.C. During the Iraq war, he suggested that UFOs being seen across the Middle Eastern region were "very likely related to Stargate/energy portal activity in Iraq/Iran and the region generally. It is very likely that the whole region comprising Iraq/Iran/Afghanistan is a vast energy portal that was strategically chosen for this reason as the home base for extraterrestrials known as the Anunnaki during the Sumerian era." The anger of Moslems over the U.S. occupation of Iraq is so intense that, according to Salla, "What they might manifest by all this rage/anger is an opening of the portals and literally the gates of hell opening with the 'return of the Gods' - the Anunnaki - who take the rage around the region as permission to intervene and punish US forces and their allies." But apparently President Bush and his fellow conspirators, who head up a secret "shadow government" in collusion with extraterrestrials, knew about this all along, and this was the real reason for the war in Iraq: "it wasn't Oil, Weapons of Mass Destruction, or the 'War against Terror', just a desire to be in Iraq if and when the energy portals/Stargates became active." He warns, "The ultimate result of extraterrestrial intervention responding to regional rage against the US is a military confrontation that could lead to a domineering extraterrestrial race having a major strategic toehold in human affairs."

Not long after this, Salla reported that "the program I had been developing, the Peace Ambassador Program [at American University in Washington, DC], was summarily terminated in late April, 2004," along with his teaching contract soon afterward. His first hint of difficulty with the University came when he was interviewed by the *Washington Post* (February 18, 2004) and told them about President Eisenhower's alleged meeting with extraterrestrials in 1954, a meeting of which the Eisenhower Library has no record. Salla characterizes his dismissal as "an example of censorship and intimidation of academic researchers exploring extraterrestrial related topics," while others applauded the university's recovery of common sense.

## Nazi Saucers and Antigravity

For decades there have been rumors about alleged "Nazi flying saucers," supposedly developed during the closing days of World War II, but too late to turn the tide of the war. In some versions of the story, Hitler and some of his top officers used these saucers to escape to a secret base somewhere in the Antarctic, from which they have been plotting a comeback to again try to take over the world. Never has there been anything resembling actual *proof* of Nazi saucers, just claims and rumors.

Salla gave a very bizarre talk at the 2005 *International UFO Congress* about the supposed Nazi-ET connection. Hitler's SS, you see, "may have" established contact with extraterrestrials, using mediums trained by esoteric secret societies like Thule and Vril. They also may have retrieved one or more crashed saucers that apparently went down in the Black Forest. long on wild claims but very short on evidence. According to Salla, the Nazis apparently established contact with tall, blonde, blue-eyed extraterrestrials, who hail from Aldebaran. Luckily the Nazis found the Aryan Aliens first, as one shudders to think how they would have treated today's four-foot greys. The Nazis employed alien technology to build flying saucers. However, they were not able to employ these wonder-weapons before the end of the war, so as the Reich was crumbling they redeployed their ET-derived technology to secret bases in Antarctica, and possibly on the Moon (which may, by the way, be artificial), and on Mars, as well. The exiled Nazis are now said to be in full cooperation with advanced extraterrestrials, and thirsting for re-conquest Salla's idea of "research" is to read wildly speculative books about UFOs whose authors present no provable facts whatsoever, and to footnote them scrupulously.

The Nazi Saucer claims were taken up again by the well-known conspiracy author Jim Marrs. His first conspiracy book was *Crossfire: The Plot That Killed Kennedy*. That one was so successful that he followed it up with several more, claiming conspiracies about extraterrestrials, secret societies, psychic warfare, 9-11 "truth", etc. He now has written *The Rise of the Fourth Reich*, which claims that unrepentant Nazi sympathizers are secretly and steadily working to create a Fourth Reich here in the U.S.

So far this is just a standard paranoid political rant. But Marrs has more interesting yarns to spin. He talks about a rumored secret "Nazi bell" (*Die Glocke*) that may have powered the Nazi saucers. The "bell" had first been written about by UFOlogist Nick Cook in his 2001 book, *The Hunt for Zero Point* (subtitled *Inside the Classified World of Antigravity Technology*). As Marrs explains in an interview on Earthfiles, "It was simply called the bell because it was kind of a bell-shaped object. This object had an outer chamber and inner chamber and created electromagnetic energy fields that were set into rotation movement. Then one layer would rotate clockwise; the other layer would rotate counter-clockwise. In doing this, it increased the power and the efficiency of this energy field generator. In doing so, there were some very unusual properties. Around the bell, plants either blossomed and grew, or withered and died; people would die around this thing."

But that's not all The Bell reportedly did. It has been claimed to be derived from extraterrestrial technology, and has been associated with time travel, UFO sightings, etc. Marrs says, "What seems to have happened in an early day version of remote viewing, or perhaps what we would call channeling, [the Nazis] did come in contact with non-human intelligences, which this contact then provided them with clues – if not direct knowledge – of technology that was far beyond what the Allies were able to come up with."

"The Bell" has now become a hot topic among seekers of "Zero Point" free energy, anti-gravity, etc. According to Cook, the Bell was about 12 feet high, and 9 feet in diameter. It was comprised of two high-speed counter-rotating cylinders, filled with a purplish, liquid-metallic-looking substance that was supposed to be highly-radioactive, code-named 'Xerum 525'. Those scientists and technicians who worked on The Bell and who did not die of its effects were wiped out by the SS at the close of the war, and the device was removed to an unknown location. It may have been loaded aboard a Nazi submarine, and taken to a secret base in Antarctica.

Cook's *The Hunt for Zero Point* is a classic example of how to spin an exciting yarn, based on almost nothing. He visits places where it is rumored that secret UFO and antigravity research is going on, such as "Area 51," and writes about what he feels and imagines, although he discovers nothing more tangible than unsubstantiated rumors. He tells an exciting story about his hotel room near Edwards Air Force Base being broken into in the middle of the night by SWAT-like forces that threaten him with automatic weapons, then he gives us the letdown – it was just in his imagination. He writes that the Northrop Grumman B-2 Stealth Bomber is likely powered by an "electrogravitic drive system," having been assured by an aviation expert that the thrust of that aircraft's engines were insufficient to lift the B-2 off the ground. Thus, facing a conundrum similar to the famous-if-apocryphal "bumblebee flight problem," he concludes that the B-2 must employ anti-gravity technology to shed some of its weight in order to get off the ground, and speculates whether there is a switch somewhere on its instrument panel to switch over to "antigravity cruise mode."

The story of the Nazi Bell has been taken up in a big way by American Antigravity (AAG), a group dedicated to promoting devices and experiments supposedly demonstrating anti-gravity effects. I have earlier written about American Antigravity's "lifters" – devices that unquestionably do rise into the air (Psychic Vibrations, *Skeptical Inquirer*, March/ April, 2003). However, they operate by using electric charges to set up an ion wind downdraft, require connection to a large external power supply far to heavy to lift into the air, and they will not work in a vacuum.

The Nazi Bell is being touted by American Antigravity as "Einstein's antigravity," because John Dering, a physicist who researches far-out stuff, speculates that "the German WW-II research was intended to create a powerful propulsion effect by engineering application of Einstein's Unified Field Theory (UFT) equations." This is also said to be tied in with the Philadelphia Experiment, an alleged Word War II era project in which the Navy supposedly rendered a Destroyer Escort ship invisible, and possibly even teleported it to Norfolk, and back. Dering claims to have seen anti-gravity effects in which mercury *fell up* in certain unspecified industrial machinery, and speculates that the Nazi bell may have operated using mercury as a medium.

Igor Witkowski, a Polish journalist specializing in military technology, has written a book called *Truth about the Wunderwaffe* (German "Wonder-Weapons") that was one source for Cook's claims. However, if you want to read these marvelous stories, his book will set you back $80 (which, it is suggested on the AAG site, is "certainly worth it.")

Joseph P. Farrell, author of *The SS Brotherhood of the Bell*, ties in the Nazi Bell with the alleged UFO crash in Kecksburg, Pennsylvania in December of 1965, largely because of the reportedly similar shape of each. The alleged Kecksburg "UFO crash" has been definitively identified as being the same as the Great Lakes Fireball of Dec. 9, 1965, widely seen across several states, and Canada (chapter 4). However, UFOlogists have never been inclined to let a few facts ruin a good story. According to Farrell, the Kecksburg Bell-UFO was later seen at Wright-Patterson Air Force Base, which also allegedly received the bodies of the dead aliens recovered from Roswell. His book costs only $17, considerably less than Witkovski's, a bargain since each are equally vacuous when it comes to proving their speculations. Farrell suggests that the lack of real evidence for claims about The Bell is because "someone, somewhere, is using 'active measures' including so-called 'wet operations' to maintain the secrecy surrounding the technology." So if you delve too deeply into its secrets, someone, somewhere is likely to rub you out. Farrell suggests, concerning the Apollo program, that "there may have been a hidden or alternative technology involved in the Lunar Excursion Module (LEM) that got us OFF the Moon," since "I don't really see the signatures of a rocket taking off from the Moon in those films of the LEMs taking off. It doesn't look like an acceleration that is geometric enough to be a rocket; it just sort of "pops up" and off it goes at more or less – it looks to me – like uniform velocity." But he admits, "I haven't actually done any measurements or not to see if this is the case."

Yet another amazing anti-gravity is on the website of our friend Richard Hoagland. Hoagland claims to reveal "Von Braun's 50-Year-Old Secret": that when America's first satellite went into orbit in 1958, Explorer I went into a higher-than-expected orbit not because of poor control of its Jupiter C rocket engine, but because of an "anti-gravity effect" somehow operating on that satellite. However, the U.S. Government ordered an "immediate cover-up" of this discovery, and Hoagland was the first person in the past fifty years to discover it. In January of 1959, a Soviet-launched lunar probe missed the moon completely and went into a solar orbit, also supposedly because of anti-gravity effects. Two months later, the U.S.-launched Pioneer 4 also missed the moon. He ties this all in with a supposed discovery that a swinging Foucault pendulum reverses itself during a solar eclipse, defying all known laws of physics (and the eclipse need not even be total). Apparently, anti-gravity forces stopped bothering satellites launched after about 1960, supposedly because Von Braun's discoveries allowed NASA to compensate for them.

As for photos of the **Nazi saucers** themselves, while for years there were no photos to be seen – only drawings – now you can find lots of "photos" using Google Images. You can decide for yourself if they came from some recently-opened secret Nazi archive, or from Photoshop.

# President Obama Goes to Mars

The stories spun in Exopolitics kept getting unbelievably weirder and weirder – and yet they kept on coming! Andrew D. Basiago, described as a "lawyer, writer, and statesman, [who] was one of America's early time-space explorers," has teamed up with Laura Magdalene Eisenhower, the great-granddaughter of Dwight Eisenhower. They claim to reveal a "secret Mars colony" that is "funded by black budget military and intelligence sources as a survival mechanism for the human genome in the event that solar flares, nuclear war, or some other cataclysm ends human life on Earth." In fact, Basiago claims to have been teleported twice in 1981 to a secret U.S. base on Mars, a journey he neglected to mention in his earlier writings when he claimed the momentous discovery of living things on Mars in photos taken by NASA's Spirit Rover. Surely he must have picked up more information about the native Martian flora and fauna during his sojourn on Mars than can be gained from scrutinizing NASA photos.

As one might expect, the proof offered up to support this claim is pretty thin. Ms. Eisenhower claims she was "the subject of a sophisticated attempt to recruit her into the secret Mars colony in 2006 that she broke free from after resisting its deeply rooted manipulations and chose to live out her destiny on Earth." As for Basiago, he speculated that "the Mars colony is being staffed by individuals who descend from specific Aryan bloodlines that contain a Martian genetic substrate and that do not represent the genetic diversity of the entire human race on Earth." Both urge "a treaty between Earth and Mars society that would establish a Mars protectorate and normalize diplomatic recognition and immigration between the two planets." *Immigration?* From Mars? Did I miss something? Who or what is going to immigrate?

Richard Hoagland has no use for claims like these. He says about Basiago, "He's an absolute 100% nut and he is being paid by the intelligence agencies to spread disinformation far and wide." Hoagland agrees with Basiago about Mars being the home of an advanced civilization, but insists that they're all dead. "All we see are ruins... literally thousands of square miles ... of ruins. And, a lot of mud covering them ... slowly eroding and blowing away in the wind. Which is why we can now see glimpses [of] what was once buried in a vast, planetary catastrophe – Which suddenly ENDED the Martian Civilization ... a long, long time ago."

And UFOlogist Art Greenfield isn't buying any of this talk from those he calls "exo-political posers." He is the author of "*Warning*," a book telling how "Reptoid aliens have been using us as a food resource for thousands of years. They came from Earth originally.... The government knows about it but keeps quiet to avoid public panic." President Obama, who he describes as "the Repto Sapien's mindless telepromptor handpuppet," is "about to sign an arms reduction treaty with the Russians to dismantle much of our nuclear arsenal. HELLO. That would result in both the Russians and us having less nukes that we might need to defend ourselves against a hostile alien takeover." Greenfield concludes, "let's stop wasting any of your valuable mental computer time on the likes of Webre and Salla and get back to solving any problems that stop us from getting on a military parity with the Repto Sapiens. We need to negotiate with them from a position of strength to persuade them to let us supply them food from sources lower down on Earth's food chain." Let 'em eat krill.

Webre's claims have gotten so bizarre (the war between the Andromeda Council and the Reptilians, Americans being teleported to a secret base on Mars to meet with aliens) that even others in exopolitics became alarmed, and began distancing themselves from him. (This recalls the old joke about the tenor who was so stupid that even the other tenors noticed it.) As Salla wrote on his website, "Webre is a marginal and controversial figure in the network of exopolitics researchers and activists that has formed around the world. Webre's writing and behavior is seen as too bizarre and controversial for most credible exopolitics researchers to use." Nonetheless, Webre's far-out articles on *The Examiner* were being read by as many as 400,000 people each month. Until recently, that is, when *Examiner.com* gave him the boot.

Webre's colleague Jon Kelly, "a world-famous expert in the application of voice-based disclosure technology for revealing UFO secrets" (whatever that is ), writes

> Kuala Lumpur War Crimes Tribunal Judge and founder of Exopolitics Alfred Lambremont Webre is calling for consumers to occupy an immediate boycott of pay-per-impression advertising funded news

website Examiner.com.... Examiner.com's corporate publication ban against the Seattle Exopolitics Examiner is an Illuminati agenda-inspired media hit targeting the columnist who revealed President Barack Obama's participation in the CIA's secret Mars visitation program.

Pieces on Examiner.com are presented to make you think you're reading a news story. One time a colleague and I were discussing one of Webre's absurd columns in Examiner.com claiming that NASA was promoting (not debunking) fears of an "extinction level event" from Comet Elenin, and he asked me "What kind of newspaper is publishing crazy stuff like this?" I replied, "I don't think it's actually a newspaper, it's more like this guy's blog." Well, in his feud with Examiner.com, Webre spelled out Examiner.com's business model:

> During 2009 many of the writers were receiving $0.01 per page view. Examiner.com later offered a variety of pay scale options to their writers. Examiner.com now bases compensation on variables such as subscriptions, page view traffic and session length.... Examiner.com derives the bulk of its revenue from consumer (reader) click-throughs. Every time you as a reader click through to read an article on Examiner.com, the company is paid a royalty by its advertisers.

At one cent per page and 400,000 pages per month, this gets you $4,000 a month, although Webre suggests that was only in a good month. So the formula for success as a UFO writer seems to be: make up the most outrageous claims you can think of and put it on Examiner.com, then sit back and collect the coins dropping into the hopper. And here am I, stupidly wasting my time and effort writing a skeptical book and Blog! Webre does not say exactly what happened to get him kicked off Examiner.com. It cannot be that they are concerned about their journalistic credibility, for they have none. Webre does say, "Examiner.com has been criticized for its lack of verification and fact-checking of stories published on the site, including accusations of plagiarism," although he does not admit making any contribution to this problem.

Webre's request is simple: "Please let your friends and networks know you are boycotting Examiner.com because it is promoting the CIA's Obama on Mars cover-up, and its direct assault on the Truth movement and Truth movement journalists like Alfred Lambremont Webre." How would that be for a Facebook status?

## Denver Extraterrestrial Affairs Committee?

Perhaps you saw the news stories in 2010 about the "Denver initiated Ordinance 300," an initiative to "require the creation of an extraterrestrial affairs commission to help ensure the health, safety, and cultural awareness of Denver residents and visitors in relation to potential encounters or interactions with extraterrestrial intelligent beings or their vehicles, and fund such commission from grants, gifts and donations." Ordinance 300 was essentially an attempt to implement the concerns and goals of Exopolitics within the existing political framework. This proposal is even crazier than it sounds. On the website for *Yes on 300* there were many astonishing claims, such as that NASA routinely removes images of UFOs from its space photos, that the U.S. government not only covers up evidence of UFOs, but also of "clean energy technologies of extraterrestrial origin, that could replace fossil fuels." And by way of a FAQ, the measure's chief promoter, Jeff Peckman, offered a new Ebook in which an "extraterrestrial being answers fifty questions," channeled "neuro-biologically via tele-thought communication" by one Lavendar, "a master astrologer" as well as "a Pleiadean contactee, emissary and scribe."

Perhaps it will restore your faith in human reason to learn that this whacked-out measure lost big, approximately 84% against vs. just 16% in favor. Or perhaps it will cause you concern to learn that approximately one voter in six thought this "extraterrestrial affairs" twaddle had substance. I suspect, however, that many voters - especially younger ones - voted "Yes" on this measure largely as a joke. Let's hope that's all it was.

## Happy World Disclosure Day!

UFO activists declared July 8, 2011 to be the first annual "**World Disclosure Day**." That date was chosen because it's the anniversary of what they see as the beginning of the supposed Roswell crash cover-up. According to *AOL Weird News*, Stephen Bassett is "a registered lobbyist who runs the Extraterrestrial Phenomena Political Action Committee an organization that since the late 1990s has been demanding Congress release information about the presence of aliens as soon as possible." He says "There was an arms race, a space race, and now we have a disclosure race. There are a dozen or so countries that might well effect disclosure tomorrow. It is hoped the Obama administration will become aware of this and take action." However, the Big Oil companies are pressuring the government to keep secret the free energy that alien technology offers us. So here we are, mire than six decades and the U.S. government still has not "disclosed" the presence of aliens all around us.

When the Obama administration launched its "We the People" petition program they probably never expected to get a petition like this. Signers were promised that the administration would give a formal response to any petition gathering 5,000 signatures within 30 days. One of the very first petitions to qualify for a response was the so-called disclosure petition, whose title says "we petition the Obama administration to: formally acknowledge an extraterrestrial presence engaging the human race - Disclosure." The petition, written by Bassett, states:

We, the undersigned, strongly urge the President of the United States to formally acknowledge an extraterrestrial presence engaging the human race and immediately release into the public domain all files from all agencies and military services relevant to this phenomenon.

This goes well beyond a request for simple "disclosure," for the release all documents. It presupposes that the U.S. government is involved in ongoing contact with extraterrestrials, and demands that the Obama administration acknowledge that fact. However, the petition quickly gathered enough signatures to qualify, creating a dilemma for the White House. All they could do is deny that they are meeting with extraterrestrials on a daily basis, and of course the denial will further fan the conspiracy flames. For a government agency to address UFO claims in any way is an automatic no-win situation. If you say they're not a real mystery, you get a controversy over that. On the other hand, if you say that UFOs are mysterious and challenging, you stir up an even bigger controversy, and many people suspect (quite reasonably) that you have been nibbling peyote or something.

On Nov. 7, 2011 the White House finally issued its response to the Disclosure Petition, which reads, in part:

# Give Me Disclosure, or Give Me Death!

Thank you for signing the petition asking the Obama Administration to acknowledge an extraterrestrial presence here on Earth. The U.S. government has no evidence that any life exists outside our planet, or that an extraterrestrial presence has contacted or engaged any member of the human race. In addition, there is no credible information to suggest that any evidence is being hidden from the public's eye.

As far as I can tell, this response is 100% correct and accurate. However, as you might imagine, it did not go down well with Bassett. PRG states simply, "The response was unacceptable." They wrote a rebuttal to the White House, urging reporters to review "quotes by persons of high rank and station over many decades," and the typical far-out pro-UFO claims.

The White House's threshold for a petition response was then raised to 25,000, although the change was not retroactive. However, Bassett was urging his followers to raise the signature count over 25,000 anyway, as a demonstration of their strength. The final signature count was 12,078.

Stephen Basset got himself very busy, and placed the "Disclosure II Petition" on the White House website on December 1, 2011.

"We petition the Obama Administration to demand a full congressional investigation of UFO/ET Disclosure efforts by the Clinton OSTP [Office of Science and Technology Policy] - the Rockefeller Initiative."

"In response to the first Disclosure Petition, the OSTP stated, "The U.S. government has no evidence that any life exists outside our planet, or that an extraterrestrial presence has contacted or engaged any member of the human race."

"If true, what was the OSTP investigating from March 1993 to October 1996 in concert with billionaire and Clinton friend, Laurance Rockefeller?

"Those who knew of and have not spoken publicly of this initiative include: Bill Clinton, Sec. of State Hillary Clinton, Obama transition co-chair John Podesta, Sec. of Defense Leon Panetta, Dr. John Gibbons, Albert Gore and Governor Bill Richardson."

Note that the second "Disclosure" petition suggested that "Hillary Clinton" and "Albert Gore," among others, were involved with the cover-up. However, this time it needed 25,000 signatures by the end of December, which it did not get. It was a bizarre subject for a petition: to request that the Obama administration demand a congressional investigation of the Clinton administration's "ET disclosure" efforts (*Psychic Vibrations*, p. 88), as if anyone outside a few UFO zealots even remembers that minor footnote to UFO history.

Basset was not the only one petitioning for "Disclosure." UFO author Richard Dolan and Hollywood producer Bryce Zabel posted a petition to "Investigate Unidentified Aerial Phenomena as Reported by Citizens, Police, Astronauts, Pilots and the Military" on the White House website. This petition, a better-crafted one in my judgment, called on the United States government to "conduct an independent investigation. This inquiry must transparently review the key unsolved UAP reports with access to classified documents. It must have the power to call witnesses and grant immunity. The findings should be publicly presented." However, it drew even fewer signatures than Disclosure II. The White House later raised the threshold for a response to 100,000 signatures.

John Podesta, former chief of staff to president Bill Clinton, former senior adviser to President Obama, and chairman of Hillary Clinton's 2016 presidential campaign, is a true believer in UFO coverups. Podesta tweeted on February 13, 2015, "Finally, my biggest failure of 2014: Once again not securing the disclosure of the UFO files." Responding to a reporter's question at a campaign stop in Conway, New Hampshire on December 29, 2015, Clinton promised, if elected, she would "get to the bottom" of UFOs and Area 51.

"UFO Disclosure" remains a dream, but it is a dream that seems destined to go on for a very long time.

# 8. UFOS - INTERPLANETARY, OR WHAT?

To most people, UFOs are synonymous with "interplanetary spacecraft." But the more sophisticated UFO proponents realize that there are very serious problems with that kind of hypothesis (which is why we hear bizarre explanations like *interdimensional* or *ultraterrestrials*). You can't just sail around the galaxy, worse yet the entire universe, the way one might set sail on the earth's oceans for a distant port. Even if you could travel at almost the speed of light, a virtually impossible task, it would still take decades, if not centuries, to get to some interesting nearby stars. You would not perceive it as being that long, but the people back on earth certainly would. Seafaring is a very bad analogy for spacefaring.

In the early days of flying saucers, it was widely assumed that the home planet of the saucers might be Mars, or Venus. As recently as 1964, no less an authority than Arthur C. Clarke was assuring us, "We are now fairly certain that there is some sort of vegetation on Mars; the seasonal color changes, coupled with recent spectroscopic evidence, give this a high degree of probability" (Clarke 1964 p. 90). Wrong answer. We have now studied much of the surface of Mars up close and personal, and have seen no evidence whatsoever of anything alive there, at least nothing big enough for us to see. We cannot yet rule out the possibility that there may be microscopic life forms on Mars, or perhaps elsewhere in our solar system.

George Adamski (1891-1965) had implausible interplanetary friends who were said to hail from Venus, as were the little men whose alleged saucer crash in Aztec, New Mexico was written about by Frank Scully. We now know that Venus' clouds consist not of water vapor, like ours, but of sulfur dioxide. Its atmospheric pressure is more than ninety times that of Earth, and its surface temperature exceeds 800 degrees F. We have already seen at least some images of the surface of Venus (as well as of Saturn's moon Titan), and there is no sign of anything living. If you want to search for life on Venus, I would say you are wasting your time. By the early twenty-first century, we have explored enough of our solar system to say with confidence that there are no intelligent life forms there, besides ourselves. In fact, it is looking doubtful that there is even any kind of macroscopic life form in our solar system, except for those of earth. So if we want to find someplace where intelligent saucer-builders might hail from, we must take that terrifying plunge, from the relatively cozy confines of our solar system to the cold depths of interstellar space. Watch out, that first step is a big one. The distance to Alpha Centauri, the closest solar system to our own, is approximately 50,000 times the distance from Earth to Jupiter. All of the other stars we can see are farther than that, typically *much farther*, by factors of tens, thousands, millions, even billions. We can probably all agree that, within a few centuries at most, we will have exhaustively explored all of the planets and moons in our solar system. But will we be able to explore beyond that?

## If It Can Be Imagined, Can It Be Done?

We regularly hear UFOlogists claiming that, while reported UFO encounters cannot be accepted as consistent with present-day science, future science will be able to accommodate them, and so therefore we should not reject the claims. As astronomer and Project Blue Book consultant Dr. J. Allen Hynek famously said to the House Committee on Science and Astronautics in 1968,

> I cannot dismiss the UFO phenomenon with a shrug. The "hard data" cases contain frequent allusions to recurrent kinematic, geometric, and luminescent characteristics. I have begun to feel that there is a tendency in 20th-century science to forget that there will be a 21st-century science, and indeed, a 30th-century science, from which vantage points our knowledge of the universe may appear quite different. We suffer perhaps, from temporal provincialism, a form of arrogance that has always irritated posterity.

Echoes of this statement are commonplace among UFO proponents. The situation is further confused by Arthur C. Clarke's famous statement that "Any sufficiently advanced technology is indistinguishable from magic," which people interpret to mean "reports of something that seems to be magic must be an example of an advanced technology." (Clarke made some other really loopy predictions about future breakthroughs, like how wheels and roads will soon be obsolete because we'll all be riding in hover cars.)

A landed UFO is alleged to simply take off from the ground and zoom away, without expelling anything in the opposite direction. Momentum has been created - how? The UFO has acquired kinetic energy as it speeds away. Where did that energy come from? Magic, perhaps? So it would appear that "future science" will no longer be limited by simplistic concepts such as conservation of energy or momentum. Even many skeptics fall into this trap. Once I was being interviewed by a well-known skeptic for a podcast, who suggested that 'before long, our technology will be able to do the things that these UFOs are reportedly doing.' And I replied that's not true, unless you are willing to toss out fundamental physical laws.

I was very interested to read in the *San Diego Union Tribune* (December 2, 2012) a story by science reporter Gary Robbins titled "Flying cars and teleporters aren't in your future," based upon an interview with UCSD physics professor Tom Murphy. Murphy relates how one day when he was talking with a group of physics students, one of them said, "If it can be imagined, it can be done." Other students nodded their heads in agreement. Said Murphy, "It took me all of two seconds to violate this dictum as I imagined myself jumping straight up to the Moon... I wondered how pervasive this attitude was among physics students and faculty. So I put together a survey. The overriding theme: experts say don't count on a Star Trek future."

Murphy designed a survey on Futuristic Physics to determine physicists' expectations of the likelihood of hypothetical future breakthroughs. The details are in his Blog *Do The Math*. One, "autopilot cars," already exists today: Google has built them, and they seem to work tolerably well. But the survey asks about a lot of other things: practical personal jetpacks; a flying car; teleportation; warp drive; wormhole travel; visiting a black hole; artificial gravity; time travel, etc. Estimates were solicited from physics undergrads, physics grad students, and physics professors. For each "breakthrough," survey participants were asked to choose one of six answers, from "likely within 50 years" to "<1% likely to ever happen, or impossible."

# Bad UFOs

As might be expected, undergrads are the most optimistic about future "breakthroughs," grad students less so, and physics professors the most pessimistic of all. It seems that the more you know about physics, the less likely you are to accept the far-out stuff. However there was one dissenting faculty member:

> Note the optimistic outlier in the faculty ranks. We saw this individual stand out on the wormhole question. Examining this person's responses, it's all 1, 2, and 3 responses, save one 4 for time travel. Nothing is off limits to this professor, and most things deserve a timescale. This individual is clearly out of step with the cohort, and tying the most optimistic undergrad: forever young.

Participation in the survey was anonymous for invited persons, but if I had to take a wild guess, I'd say that Prof. Michio Kaku probably participated. (Kaku praised Leslie Kean's problem-ridden UFO book as the "gold standard" of UFO research.) Murphy notes,

> The biggest differences between faculty and grad students crop up on questions pertaining to flying cars, cloaking, and studying astrophysical objects up close. The largest graduate-undergraduate discrepancy appears for the question about artificial gravity. The largest end-to-end discrepancies (faculty to undergraduate) relate to flying cars, artificial gravity, and warp drive.

The physics faculty members' expectations of the likelihood of certain developments, from most to least probable, is as follows:

| | |
|---|---|
| Autopilot Cars | likely within 50 years |
| Real Robots | likely within 500 years |
| Fusion Power | likely within 500 years |
| Lunar Colony | likely within 5000 years |
| Cloaking Devices | likely within 5000 years |
| 200 Year Lifetime | maybe within 5000 years |
| Martian Colony | probably eventually (>5000 yr) |
| Terraforming | probably eventually (> 5000 yr) |
| Alien Dialog | probably eventually (> 5000 yr) |
| Alien Visit | on the fence |
| Jetpack | unlikely ever |
| Synthesized Food | unlikely ever |

| | |
|---|---|
| Roving [close-up] Astrophysics | unlikely ever |
| Flying "Cars" | unlikely ever |
| Visit Black Hole | forget about it |
| Artificial Gravity | forget about it |
| Teleportation | forget about it |
| Warp Drive | forget about it |
| Wormhole Travel | forget about it |
| Time Travel | forget about it |

So to those who are proclaiming that UFOs are real, and that 'future physics' will explain how they operate via wormholes, warp drives, teleportation, or time travel, the message from physics professors is: *forget about it*.

## How Plausible is Interstellar Travel?

**Stanton Friedman**, the "Flying Saucer Physicist," is confident that interstellar travel is not only possible, but fairly likely. In his essay *UFO Propulsion Systems*, Friedman writes,

> "a one-way trip of thirty-seven years (the distance to Zeta 1 or 2 Reticuli) at 99.9 percent c would take only twenty months' crew time; at 99.99 percent c it would take only six months' crew time. Thus even a trip to a distant galaxy such as Andromeda, two million light-years away, would take under sixty years' crew time if the intergalactic ship somehow could manage to keep accelerating at one G, using some yet unknown technique."

Ah, that pesky little "yet unknown technique." Now this is all perfectly true, but it blithely ignores some very fundamental problems that are *not related to any level of technology*. A trio of "classic" papers written in the 1960s by physicists examine the fundamental physics involved in proposed interstellar travel, and explain the formidable obstacles: **obstacles imposed by fundamental laws of physics, not by limits of technology**. These articles sufficed to convince the scientific community that the concept of interstellar travel is utterly implausible, and explanations for UFO sightings must be sought elsewhere, in psychology and sociology, not in physics. However, in recent years these articles have largely been overlooked, so I think it's very important to examine each one carefully:

**1. *Radioastronomy and Communication Through Space*** by Edward Purcell. (U.S. Atomic Energy Commission Report BNL-658, reprinted in Cameron 1963.) Purcell (1912-1997) was in the physics department at Harvard University, and shared in the 1952 Nobel Prize for physics. He was a pioneer in radio astronomy, the first to detect the famous 21-cm radio emission line from neutral hydrogen in the galaxy. He also is credited with the discovery of Nuclear Magnetic Resonance, used in MRI machines today.

Most of the paper is uncontroversial and explains then-recent discoveries in radio astronomy. But in the section titled *Space Travel*, Purcell examines claims that someday we will travel to the stars at almost the speed of light. "The performance of a rocket depends almost entirely on the velocity with which the propellant is exhausted," he notes. Thus, "the elementary laws of mechanics – in this case relativistic mechanics, but still the elementary laws of mechanics – inexorably impose a certain relation between the initial mass and the final mass of the rocket in the *ideal* case... **It follows very simply from conservation of momentum and energy, the mass-energy relation, and nothing else.**" [emphasis added]

> "For our vehicle we shall clearly want a propellant with a *very* high exhaust velocity. Putting all practical questions aside, I propose, in my first design, to use the *ideal nuclear fusion* propellant... I am going to burn hydrogen to helium with 100 percent efficiency; by means unspecified I shall throw the helium out the back with kinetic energy, as seen from the rocket, equivalent to the entire mass change. You can't beat that, with fusion. One can easily work out the exhaust velocity; it is about 1/8 the velocity of light. The equation of Figure 13 tells us that to attain a speed 0.99c we need an initial mass which is a little over a *billion* times the final mass."

*A billion times the final mass*?????!!!!!!! In fact, the exact figure is 1.6 X $10^9$. So in the ideal case, where you had somehow mastered nuclear fusion with 100% efficiency and could control and direct the energy in whatever way you choose, you still will need ***1.6 billion tons of fuel*** for each ton of payload! Surely, such a rocket has never been built, and never will be built, in our solar system, or any other. Thus Purcell has demonstrated, beyond any possibility of doubt, that all proposals to reach near-light speed using nuclear fusion propulsion are complete absurdity.

# Interplanetary UFOs?

But supposing some other, more energetic reaction could be found? Nuclear fission produces an even lower exhaust velocity than fusion, so it's even less plausible. Is there any reaction more energetic than nuclear fusion?

> This is no place for timidity, so let us take the ultimate step and switch to the perfect matter-antimatter propellant.... The resulting energy leaves our rocket with an exhaust velocity of $c$ or thereabouts. This makes the situation very much better. To get up to 99 percent the velocity of light only a ratio of 14 is needed between the initial mass and the final mass.

That sounds very much better. If I can "somehow" procure sufficient antimatter, "somehow" store it, and "somehow" control its reaction with matter, and "somehow" direct the resulting energy in the direction opposite where I want it to go, I need only 7 tons of matter, and 7 tons of antimatter for each ton of payload. That sounds almost possible. But Purcell points out that all that buys you is a one-way ticket out of the galaxy: you have no way to slow down and stop when you get where you want to go. So to stop when you reach your destination requires that fuel-to-payload ratio to be squared: 196. And if you want to someday return, unless you know of a convenient matter-antimatter fueling station at your destination, you will need to square that again, for a fuel-to-mass ration of almost 40,000.

And even if you could "somehow" construct such a vehicle, your problems are not over. "If you are moving with 99 per cent the velocity of light through our galaxy, which contains one hydrogen atom per cubic centimeter even in the 'empty spaces," each of these hydrogen atoms looks *to you* like a 6-billion-volt proton, and they are coming at you with a current which is roughly equivalent to 300 cosmotrons per square meter. So you have a minor shielding problem to get over before you start working on the shielding problem connected with the rocket engine." [Remember that Friedman proposed that UFOs might be spaceships traveling at 99.99% the speed of light, not just 99%, so these problems become much worse.] Also, "In order to achieve the required acceleration our rocket, near the beginning of its journey will have to radiate about $10^{18}$ watts. This is a little more than the total power the earth receives from the sun. But this isn't sunshine, it's gamma rays. So the problem is not to shield the payload, the problem is to shield *the earth*."

"Well, this is preposterous, you are saying. That is exactly my point. It *is* preposterous. And remember, our conclusions are forced on us by the elementary laws of mechanics." ***Nothing else needs to be written about the possibility of relativistic interstellar travel – Dr. Purcell has shown it to be completely preposterous.*** Purcell concludes his paper, however, by demonstrating that interstellar communication using radio waves is perfectly possible. His article closes with, "All this stuff about traveling around the universe in space suits – except for local exploration, which I have not discussed – belongs back where it came from, on the cereal box."

**2. *The General Limits of Space Travel*** by Sebastian von Hoerner (*Science* 137, 18, 1962; reprinted in Cameron 1963). Immediately following Purcell's paper in the Cameron volume is this related paper by von Hoerner (1919-2003), a German radio astronomer who was influential in early discussions and proposals for SETI. He examines the physical difficulties of propulsion for space travel, including possibilities not covered by Purcell. Von Hoerner considers ion thrust propulsion, but concludes that "nuclear reactors and all the equipment needed to give a strong ion thrust are so complicated and massive, as compared with the relatively simple combustion equipment, that there is no hope at present of reaching, with reactors, the value of $P$ [engine power to mass ratio] already attained with combustion rockets." He also considers proposals for a huge "scoop" or funnel for a rocket to fuel itself as it goes along, scooping up galactic hydrogen. But he notes that interstellar matter has very low density, and "in order to collect 1000 tons of matter (10 times the fuel of one Atlas rocket) on a trip to a goal 5.6 parsecs away, one would need a funnel 100 km in diameter; we will rule out this possibility."

After several pages of equations covering much the same ground as Purcell, Von Hoerner concludes, "there is no way of avoiding these demands [for power], and definitely no hope of fulfilling them...space travel, even in the most distant future, will be confined completely to our own planetary system, and a similar conclusion will hold for any other civilization, no matter how advanced it may be. The only means of communication between different civilizations thus seems to be electro-magnetic signals."

**3. *Physics and Metaphysics of Unidentified Flying Objects*** by William Markowitz (*Science* 157, 1274, 1967). Markowitz (1907-1998) was an

# Interplanetary UFOs?

Austrian-born astronomer who worked at the U.S. Naval Observatory, and also taught astronomy and physics at Pennsylvania State University and Marquette University. He was a pioneer in the use of atomic clocks for astronomy, specializing in precision time measurement issues. Markowitz wrote, "Aristotle wrote on natural phenomena under the heading 'physics' and continued with another section called 'metaphysics' or 'beyond physics.' I use a similar approach here. First I consider the physics of UFO's when the laws of physics are obeyed. After that I consider the case where the laws of physics are not obeyed. The specific question to be studied is whether UFO's are under extraterrestrial control." By the laws of physics, he is concerned with only the simplest and best-known ones, like those of motion, gravitation, conservation of energy, and the restrictions of special relativity. He points out an obvious but seldom-noted problem: "Apart from propeller and balloon action, a spacecraft can generate thrust only by expelling mass." And something that uses propellers or balloons is an *aircraft*, not a spacecraft.

UFOs are sometimes reported to land, and take off again.

> "If an extraterrestrial spacecraft is to land nondestructively and then lift off, it must be able to develop a thrust slightly less than its weight on landing... if nuclear energy is used to generate thrust, then searing of the ground at 85,000 deg C should result, and nuclear decay production equivalent in quantity to those produced by an atomic bomb should be detected. This has not happened. Hence, the published reports of landing and lift-offs of UFO's are not reports of spacecraft controlled by extraterrestrial beings, if the laws of physics are valid."

> "We can reconcile UFO reports with extraterrestrial control by assigning various magic properties to extraterrestrial beings. These include 'teleportation' (the instantaneous movement of material bodies between planets and stars), the creation of 'force-fields' to drive space ships, and propulsion without reaction. The last of these would permit a man to lift himself by his bootstraps. Anyone who wishes is free to accept such magic properties, but I cannot."

To those who were following the controversy at that time over the proposal championed by J. Allen Hynek and Jacques Vallee for a "scientific study of UFOs," an 'ulterior motive' for the Markowitz article was immediately apparent. The previous year Hynek had a letter published in *Science*, arguing that UFOs were worthy of scientific study (*Science* 154, 329, 1966). Markowitz carefully notes several instances where Hynek and his colleagues contradicted themselves in their statements about UFOs. For example, in his letter in *Science*, Hynek wrote, "Some of the very best, most coherent reports have come from scientifically trained people." But Markowitz noted that Hynek had written quite the opposite in his article in the *Encyclopedia Britannica* two years earlier: "It appears unreasonable that spacecraft should announce themselves to casual observers while craftily avoiding detection by trained observers." Markowitz further noted that Vallee's 1966 book *Challenge to Science* presents the "classic" 1948 sighting of pilots Chiles and Whitted, who reported a dramatic close encounter with a huge metallic object while flying a DC-3; "the book fails to mention that Hynek had identified the object as an undoubted meteor in his report of 30 April 1949 to the Air Force... This omission is curious because Hynek wrote a foreword to *Challenge to Science*." These and other self-contradictions, compiled by Markowitz, showed that the Hynek/Vallee case for the UFO was utterly lacking in intellectual rigor. It unmasked the real Hynek, disorganized, indecisive, and confused. This paper, published in the peer-reviewed pages of *Science*, was fatal to the credibility of Hynek's proposed "scientific study of UFOs." There were, and still are, a few scientists who took Hynek's UFO theorizing seriously, but they have always been a very small group.

## What About "Wormholes"?

Some theorists of interstellar travel are quite aware of the extreme difficulties involved in actually traveling to interstellar destinations, in the sense of going from Point A to Point B. So they hypothesize easier ways to reach interstellar destinations, without the pesky problem of traversing every point between them. Maybe we can warp space so that the distance between earth and the Andromeda galaxy is not two million light years, as in ordinary space travel, but far, far less? Suppose there is a wormhole with one end where we now are, and the other where we want to go?

The "Bohemian physicist" Dr. Jack Sarfatti of San Francisco is a colorful figure. He has written papers claiming that wormholes can be used not only to travel through space, but through time as well. He suggests that

# Interplanetary UFOs?

UFOs are real, and travel through wormholes to reach us from some other place or time.

According to Wikipedia,

> "**Wormholes** which could actually be crossed, known as traversable wormholes, would only be possible if exotic matter with negative energy density could be used to stabilize them. (Many physicists such as Stephen Hawking, Kip Thorne, and others believe that the Casimir effect is evidence that negative energy densities are possible in nature.) Physicists have not found any natural process which would be predicted to form a wormhole naturally in the context of general relativity, although the quantum foam hypothesis is sometimes used to suggest that tiny wormholes might appear and disappear spontaneously at the Planck scale, and stable versions of such wormholes have been suggested as dark matter candidates. It has also been proposed that if a tiny wormhole held open by a negative-mass cosmic string had appeared around the time of the Big Bang, it could have been inflated to macroscopic size by cosmic inflation.

So yes, a wormhole is something that might theoretically exist, although their actual existence is extremely dubious. There is no reason to think that they could occur naturally, and no observational evidence that they actually do exist (unlike Black Holes). Even if wormholes do exist, they may exist only on the Planck scale (subatomic quantum size). It seems extremely dubious that traversable wormholes exist in nature, and even if they do, we still have seemingly insurmountable problems. How do we find wormholes? How do we determine whether they are stable? How do we know where their destination is? If we go into one, is it possible to return? There is also the problem of simply getting to the wormhole's mouth. If the wormhole were near our solar system, we would already detect its disturbing effects of warped space. And if it is far from our solar system, we need to develop interstellar travel simply to travel to the wormhole's mouth!

Can we create a wormhole to go from where we are to where we want to be? Perhaps in theory we might, but the reality of a recipe for creating a wormhole will undoubtedly be something like this:

Take 100 solar masses. Bake at one million degrees for ten thousand years. Stir in 50 solar masses of exotic matter with negative energy density. Stretch out the mix from desired source to destination. Let cool for one million years.

So the idea of using wormholes as a convenient transportation network to wherever in the universe we want to go is, well, fanciful and implausible in the extreme. We can't proclaim it completely "impossible," but the person who proclaims it as a reality had better have extraordinarily good evidence that such a thing exists.

# UFOs From Fairyland

After UFOs had been around for about twenty years, a few UFO theorists began to realize that the common "extraterrestrial hypothesis" really made no sense. Not only are the difficulties of interstellar travel obvious and overwhelming, but UFOs often are reported to operate in a way that no solid- 3-dimensional object could, appearing or disappearing in ways that seem to defy all laws of physics. Plus, they seem to have a strange property of selective visibility, what I have termed a *Jealous Phenomenon* (suspicious or watchful in terms of letting itself be seen) that also violates all ideas of common sense. A UFO might fly over New York City or Los Angeles, or even land near there, yet be seen only by one person! However, instead of concluding that the reports are flawed, some UFOlogists have concluded that UFOs must represent something even more bizarre than extraterrestrial visitors.

In 1969 Jacques Vallee wrote *Passport to Magonia: From Folklore to Flying Saucers*. In it he compares contemporary reports of UFOs and their supposed occupants to accounts of fairyland (Magonia) and other traditional legends. When I first heard about the book, I thought "Well, Vallee has finally come to his senses, and realized that UFO stories are just the same as old folklore and legends, only in modern garb." But when I read the book, I realized how wrong I was: Vallee was arguing that UFOs' resemblance to the legends of old deepens the mystery of both!

Vallee does not shy away from claims about earthlings having sex with aliens, comparing it to earlier reports about sex involving incubi, succubae, angels, and demons. "What we have here is *a complete theory of contact between our race and another race, non-human*, different in physical nature, but biologically compatible with us. Angels, demons, fairies, creatures from heaven, hell, or Magonia: they inspire our strangest dreams, shape our destinies, steal our desires." We will figure out what UFO occupants really represent, suggested Vallee, as soon as we can figure out these others.

However, unlike some writers who dabble in folklore and UFOs, the British folklorist David Clarke understands that when stories parallel folklore, it suggests they are fiction, not truth. "Folklorists categorize invented or false stories as examples of what they call 'ostension', whereby people either spontaneously or deliberately create new stories and events that reproduce existing lore and legends" (Clarke, 2015). He presents a detailed argument showing how pre-1947 science fiction stories already contained most of the significant themes later turning up as 'actual' UFO claims:

> "Ockham's razor suggests that aliens are not monsters from the Id, or expressions of a collective unconscious, let alone travellers from another world. They are products of our imagination and the science fiction that has been clogging our brains for more than a century."

John Keel (1930-2009) was another UFO theorist who wrote about supposed UFO occupants as *ultraterrestrials*. In *UFOs – Operation Trojan Horse* (1970), he wrote, "Many flying saucers seem to be nothing more than a disguise for some hidden phenomenon. They are like Trojan horses descending into our forests and farmfields, promising salvation and offering us the splendor of some great supercivilization in the sky." Keel concluded, "The flying saucers do not come from some Buck Rogers-type civilization on some distant planet. They are our next-door neighbors, part of another space-time continuum where life, matter, and energy are radically different from ours." Other UFOlogists have dabbled with similar theories. Brad Steiger wrote *The Gods of Aquarius* (1976), in which he describes his own Close Encounter with an elf, who he described as "my multidimensional friend." Jerome Clark suggested in *The Unidentified* (1975) that the famous **Cottingley Fairies** hoax photos were evidence of this hidden world. UFO chronicler and satirist James Moseley (1931-2012) always said, only partly in jest, that he believed in the "3 ½ dimensional theory" of UFOs. They represented something real, he believed, but they were not entirely a phenomenon of the ordinary 3-dimensional universe in which we live.

# The Prime UFO Fallacy

When I first started to give lectures or interviews about UFOs, something surprised me. After I would explain why the UFO reports and "evidence" presented do not constitute proof that there are any extraordinary objects flying around in earth's skies, almost invariably someone would angrily object, "What makes you so certain that we're the only intelligent beings in the whole universe?" I call this *the Prime UFO Fallacy*.

Of course, I reply, "I never said that." Yet they think that, implicitly, I did. They seem to have the notion that, *if extraterrestrial intelligent beings exist, then they must be here now.* When stated so directly, the fallacy is obvious. Yet it seems to have a certain plausibility to many people. After all, we can send astronauts to the moon. Therefore other planets should be able to send their astronauts to us.

If the planets we're talking about were as close as Venus or Mars, let alone our moon, then this might make sense. But they are not. The universe is so incomprehensibly vast, and contains an unfathomable number of solar systems with planets similar to our Sun's, that even if the development of intelligent life is extremely rare, that would still imply a large number of intelligent civilizations. But in light of the difficulties we have discussed concerning interstellar travel, it is entirely possible that even if our galaxy possesses many intelligent civilizations, with distances between them measured in the hundreds if not thousands of light years, each would be isolated in its own little corner of the galaxy, effectively alone forever. However, the possibility of eventually discovering, and even communicating with, other intelligent civilizations using electromagnetic waves offers us at least some hope of overcoming our own galactic isolation and learning about some distant others.

# 9 COSMIC DOOMSDAYS

After you've been following UFO claims for a while, one theme that keeps popping up is that some sort of 'cosmic doomsday' is always just around the corner. Sometimes the claim directly involves aliens, other times it involves supposed 'cosmic forces' that are conspiring to wreak havoc on our poor little planet, always threatening great destruction, wailing, and gnashing of teeth. Not all of these Doomsdays are alien or space-related, but they are included to show the human penchant to expect – indeed, to almost wish for – the End of the World. Another common element: the supposed danger is always 100% totally and utterly bogus. There is never the slightest rational reason to be concerned about the thing that is predicted. Yet no sooner does one promised Doomsday come and go with no effect than another is proclaimed, and the whole cycle of panic begins anew. There must be some deep-seated human need to hear, and at least halfway believe, frenzied proclamations of our impending doom.

## The Jupiter Effect, 1982

In 1974, the book *The Jupiter Effect* by John Gribbin, PhD and Stephen Plagemann was published. It became a huge best-seller. Its argument was simple: in March of 1982 there will be an "alignment" when seven planets (six if we correctly ignore Pluto) will be lined up on the same side of the Sun. And "lined up" is used in the rough sense: they're within an arc of 95 degrees, or in other words, all within one quadrant, more or less. Somehow this was going to affect the Sun, which would then affect the Earth by the solar wind, and trigger a catastrophic earthquake along the San Andreas Fault near Los Angeles. I don't know why the solar wind wanted to single out Los Angeles and not Tokyo or Lisbon, but it clearly had sinful California in its sights, where book sales to easily-frightened flakes, fruits, and nuts could be maximized. There were other catastrophes that would occur as well. Probably about as many people were frightened about 1982 as are were later frightened about 2012.

And of course, 1982 came and went without any of the predicted catastrophes occurring. In February of 1982, these same two authors published *The Jupiter Effect Reconsidered*, which was also a best-seller. They didn't even wait for March to come and go before they published their apologia. It shows amazing *chutzpah* when one writes a best-seller, *The Sky is Falling*, then follows it up with another best-seller, *Why the Sky Didn't Fall*.

The authors claimed that, even though the alignment didn't cause any major earthquakes in 1982, it was responsible for triggering the giant eruption of Mt. St. Helens in 1980 (two years before the "alignment" had occurred). "Psychics" do this all the time, re-interpreting their failed predictions as successes by the time-honored practice of "moving the goalpost." If a "psychic" predicts A, which does not occur, but B does, the prediction will be re-interpreted to have meant B. Surprisingly, neither Gribbin nor Plagemann has yet been tarred-and-feathered for their journalistic malfeasance.

But **what was Isaac Asimov thinking** when he agreed to write the Foreword to *The Jupiter Effect*, and sound almost like he agrees with the prediction?

> And in the end, it turns out that something will happen in 1982 that just may – No, no, read it for yourself. Read it carefully and you'll find it far more fascinating than the tale of any millionaire stabbed in any library, locked or otherwise. And far more important, too – especially if you live in California.

Probably Asimov was thinking the same thing as the authors: Big Dollar Signs.

## Carried Away by the Rapture

Belief in the immanent return of Jesus Christ has been common since the earliest days of Christianity. The former NASA engineer and self-taught Biblical scholar Edgar C. Whisenant attracted many followers in the Evangelical Christian community with his confident prediction that the Rapture would occur in 1988, probably to coincide with Rosh Hashana. Over four million copies were sold, and many others distributed for free.

He confidently stated, "Only if the Bible is in error am I wrong," and we know that isn't possible. He also predicted that a nuclear war would break out between the US and the USSR on October 4, 1988, with the Final Judgment occurring in November 1995. When the world failed to end in 1988, he issued more rapture predictions for 1989, 1993, and 1994, but these did not attract as much attention.

The Rapture returned again in 1992. From the *Skeptical Inquirer*, Summer 1993, News and Comment item by Erik Vaughn, "Where's the Rapture?":

> In case you missed it, Wednesday, October 28, 1992 was the Rapture, the day faithful Christians dead or alive ascend to heaven. The Rapture kicks off the last days: the appearance of the Antichrist, Armageddon, and the Second Coming.

Lee Jang Rim, leader of the Dami Mission Church in Seoul, South Korea, spread the prophecy. Followers quit their jobs and sold or destroyed their property in anticipation, even as Rim was arrested for allegedly bilking hundreds of thousands of dollars from church members. The hysteria created a near crisis in South Korea. Members of Korean churches in South Korea, Hawaii, Los Angeles, and New York held marathon prayer vigils.

June 9, 1994 was another date on which the world ended. Various fundamentalist prognosticators somehow settled upon that date, and started warning ominously that "June ninth is coming," the day on which God would supposedly "rip sin out of the world." Warnings were issued for a major destructive earthquake somewhere along the Pacific "ring of fire," supposedly only the first event in an eschatological progression. So widespread was the expectation that it was even mentioned on *The 700 Club*, evangelist Pat Robertson's popular TV program, although with the clear disclaimer that it would not be the start of the Rapture. When a large underground earthquake, which caused little or no damage or casualties, struck on this date, it was hailed as fulfillment of the prophecy. The last we checked, the world still had plenty of sin.

Another Rapture was scheduled in September of 1994 by Harold Camping (1921-2013), president and general manager of a 39-station network of religious radio broadcasters. He proclaimed that before the end of that month, the dead would be arising from their graves for their final judgment. Camping explained to the *San Jose Mercury News* (Sept. 4, 1994) that he had arrived at his prediction by a laborious and careful process of counting off significant dates in the Bible. Furthermore, he boasted that his math had been checked by a nuclear physicist and "a number of scientists at the Lawrence Livermore lab," as if errors in arithmetic were the only way such a prediction might go awry. Mainstream religious scholars, of course, suggested Camping was seriously mistaken. Fortunately, Camping exhorted his listeners, "Don't do anything bizarre. Just live the way you should have been living all the time." After the non-Rapture, his stations remained on the air, and apparently found something else to talk about.

But the world was not safe from the eschaton merely because it survived into 1995. Dr. Leland Jensen of Missoula, Montana said that *he* is the Second Coming, and that during 1995 the earth would suffer great meteor impacts, earthquakes, and major planetary changes. Those lucky enough to survive all this will enjoy Heaven on Earth. And according to a story in the *Washington Post* (March 12, 1994), followers of the Institute of Divine Metaphysical Research expected the world to end in an instant, by 1996. Of course, even if these people had been right, they would not be in a position to gloat about it.

## 1997: Comet Hale Bopp Brings Closure to Heaven's Gate

One day in April of 1976, some enigmatic handwritten signs suddenly began appearing around the University of Maryland in College Park, which I was fortunate to spot. At that time I was living in the Maryland suburbs of Washington, DC, and my then-wife was earning a Masters Degree at the University of Maryland. The signs proclaimed a lecture on the evening of April 19 "to explain the UFO Two." I suspected that this mysterious meeting may involve "Bo and Peep" - Marshall Herff Applewhite and Bonnie Lu Nettles (also known as "The Two," "Him and Her," and many other names), who had recently been in the headlines as a kind of 'Pied Pipers' of UFOlogy. In numerous cities they had appeared without advance notice to give lectures about a type of "salvation" involving UFOs. They had somehow lured dozens of people away from their homes and normal lives into a UFO cult. I made plans to attend.

# Cosmic Doomsdays

It indeed did turn out to be one of the cult's recruiting meetings. Arriving early, I recognized from news photos Applewhite and Nettles standing around and chatting with the cult members. I said nothing. During the meeting, Applewhite and Nettles sat incognito among the audience while the cult members at the speakers' table talked glowingly about the coming "harvest." The UFOs would take those who were ready to be "harvested" up to the "next level", where they would live a better life. Bo and Peep are the only two people now on earth representing that higher level. The cultists on the panel obviously believed every word of this nonsense. When asked about the whereabouts of their leaders, the cultists claimed to not know where they were. "We believe they are in the Midwest somewhere." They were lying. The Two were seated in the audience, although amazingly nobody seemed to realize this. Some of the audience members were quite angry, presumably having had friends or relatives disappear into the cult - probably this is why Bo and Peep preferred to remain incognito. When I had a chance to ask a question, I raised the issue of The Two's previous brushes with the law - news reports had mentioned several - and I asked if these were the kind of persons whose word they would trust so completely. As I was speaking, Applewhite rose up from his chair on the other side of the aisle, stood full up and glared at me, from about fifteen feet away. He was a large man, and I was surprised to see that he had an air of being dangerous (which later events fully confirmed). It would have been easy to blow apart the charade by confronting him right there, but I did not. I have always regretted my failure to act in that moment, most especially in light of what ultimately happened.

Bo and Peep gradually faded from sight. A few articles and books were written, but their cult was largely forgotten, until the astonishing news burst upon the world in March of 1997. Believing bogus reports on *Coast to Coast AM* that the brilliant Comet Hale-Bopp was being followed a mysterious UFO "companion," **Applewhite** proclaimed to his followers that the sign had at last come for them to move on to the "next level," and join Nettles, who had died twelve years earlier. In Rancho Santa Fe, California, thirty-nine members of Applewhite's cult now calling itself **"Heavens Gate,"** put on their sneakers, took fatal doses of drugs and alcohol, then lay down with plastic bags over their heads expecting to "rise up" to the "next level" and join Nettles on the comet.

## 1998: God Arrives in a Flying Saucer

Approximately 150 members of a Taiwanese UFO cult called Chen Tao (Perfect Way) came to the US in 1996 expecting to meet God, who would arrive in a flying saucer. They first settled in San Dimas, California, a state where cults seem almost normal. Then suddenly they moved to Garland, Texas, the sort of town where a large crowd of foreign-speaking cult weirdos would not exactly pass unnoticed. A group of homeowners in Garland wrote the mayor and City Council complaining that this bizarre group was ruining their neighborhood. The mayor replied that these people were simply exercising their rights to free speech, and there wasn't anything he could do. However, what he should have replied was, "Don't worry, these people won't bother you for long. They say they're going to be carried away on flying saucers on the morning of March 31." For according to the group's leader Hon-Ming Chen, God was going to arrive at his house on that date in 1998 at 10:00 AM Central Time with a whole fleet of flying saucers, swooping all of them up from the earth to rescue them from the coming Great Tribulation. Authorities, expecting that the saucer fleet might not materialize, feared another mass-suicide like the Heavens Gate UFO cult the previous year, especially since many members carried neatly-packed backpacks with identical white clothing and sneakers for their anticipated heavenly rendezvous. The Taiwanese media had been reporting that Chen was encouraging newcomers to kill themselves so their bodies could be picked up by flying saucers. However, the church members denied having any thoughts of suicide.

Large numbers of reporters, in addition to Chen's followers, gathered at the appointed place and time but, alas, nothing happened. Chen had an explanation: God had already arrived, invisibly, and entered their souls. As soon as things quieted down a bit, Chen announced that God would not be gathering people up from Texas after all, but instead from the area of the Great Lakes. He and a small group of the faithful set off for Buffalo, New York. Many of the others returned to Taiwan. While in New York, Chen had a vision of the numbers 17 and 18, which he interpreted as a divine instruction to move his followers to the town of Olcott, where those two routes intersect. Then Chen prognosticated that Armageddon would occur in August of 1999. When that, too, failed to happen, the group seems to have dissolved.

## 2000 – The Y2K Bug Shuts Down Civilization

Some say the world will end by ice, others say by fire, and still others say by computer glitches. There had long been the general expectation that something absolutely dreadful would happen in the millennial year 2000 - the Rev. Louis Farrakhan foresaw a plague of earthquakes, hailstorms and floods. Until the late 1990s doomsayers couldn't agree on just was going to do us in. But suddenly most doomsayers agreed that the Year 2000 Problem for computers was the menace that would bring about the end of civilization. Usually called the "Y2K" or "millennium" bug, it was the result of the short-sighted programming technique of employing only two digits to represent a year, presupposing that the first two digits would always be "19". No one denies that the Y2K problem could have been serious, and the computer industry spent a lot of money to repair it. I was a software engineer in the Silicon Valley during the 1990s, and our department held a number of high-level meetings where we carefully examined each and every software package we relied upon to ensure that it would function after January 1, 2000, and if not, replace it. I knew that not a single one of the embedded system products that I had worked on – the kind that would supposedly be the most affected by Y2K – knew, or even cared, what the year was. They had clocks, but they were just interval timers, to ensure that real-time processing events occurred when they were supposed to. The year could have been 2000, or 9999, or -500, it would not matter.

However, according to some people it was already too late, and it was time to head for the hills. The Sept. 15, 1998 issue of the *Weekly World News* proclaimed January 1, 2000 as "THE DAY THE EARTH WILL STAND STILL." "All Banks will fail! Food Supplies Will be Depleted! Electricity Will be Cut Off! The Stock Market Will Crash! Vehicles Using Computer Chips Will Stop Dead! Telephones Will Cease to Function! Domino Effect Will Cause A Worldwide Depression!" Given that this publication has, over the years, rarely allowed any snippets of nonfiction to intrude upon its wonderfully made-up stories, one might be forgiven for assuming that their Y2K crash-and-burn scenario was as bogus as their story about soap made with holy water that revives the dead. But a lot of people took such cries of an impending Cyber-Armageddon very seriously indeed.

Longtime gloom-and-doom author Gary North devoted his entire website to the year 2000, which he called "the year the earth stands still." This is the same author who in *None Dare Call it Witchcraft* warned us that UFO and paranormal manifestations are indeed real, but Demonic in origin. The Y2K problem, he warns, "may be the biggest problem that the modern world has ever faced. I think it is. At 12 midnight on January 1, 2000 (a Saturday morning), most of the world's mainframe computers will either shut down or begin spewing out bad data...Think of what happens if the following areas go down and stay down for months or even years: banks, railroads, public utilities, telephone lines, military communications, and financial markets. What about Social Security and Medicare?" The result, says North, will be nothing less than catastrophic: "I think the division of labor will collapse in 2000. If the power grid goes completely down, it will stay down. The division of labor will collapse to early 19th century levels, except that we have lost early 19th century skills."

For those who were so scared that they literally wanted to head for the hills, entrepreneurs were busily getting the hills are ready for them. A development called Heritage Farms warned of "The Millennium. The Y2K Computer Meltdown. Economic Recession/Depression. Doomsayers predict nothing short of total collapse within the next two years. Will any of it happen? We don't know, and we seriously hope NOT, but the mounting evidence was convincing enough to make us look for a place to ride out the turmoil. We have now found it in ARIZONA!" Located 180 miles northeast of Phoenix, Arizona, Heritage Farms promised that its "capability of total self-sufficiency, and independence from outside energy sources" made it "the model rural village for what may be a whole new way of life in the first part of the next century." "500 families of the New Millennium can grow their own food and food for their neighbors to purchase or barter. They will harvest electrical energy from the sun and wind." Their original plan for a Y2K refuge in Sully County, South Dakota was unanimously voted down by local zoning officials, who feared that a Doomsday Cult might be forming in their midst. Kevin Poulsen of Ziff-Davis TV said, "as planned, the Y2K village would be one half refuge from millennial madness, and one half Disneyland. Main Street would be clean and neat with an Old West theme." A splendid place, no doubt, to stand by and watch civilization collapse.

For those not quite ready to head for the hills but wanting to stockpile food for anticipated Y2K chaos, there were many "survivalist" Y2K sites. They offered bulk rations of long-storage freeze-dried food to help ensure your survival for however many months or years it may take for your grocery store to get their cash registers back online in the new millennium. There was even a "Year 2000 Problem Site Exclusively Designed for Women," explaining "The Year 2000 Computer Problem: The 10 Things Every Woman Must Do Now to Keep Herself and Her Family Safe." Presenting gloom-and-doom survivalism with a gentle feminine touch, it contained advice on how to handle all of the expected privations of the impending millennium, even including an expected interruption in the supply of disposable products for feminine hygiene.

The technological leader of the doomsayers was a man whose judgment was, until then, seldom questioned: Ed Yourdon. A 35-year veteran of the computer industry, Yourdon attained prominence in computer science as the lead developer of the structured analysis/design methods of the 1970s, as well as a co-developer of contemporary methods of object-oriented analysis. But Yourdon, along with his wife Jennifer, wrote *Time Bomb 2000,* which became a Bible to the sky-is-falling crowd. Yourdon wrote that, "My wife and I recently sold our New York City apartment and bought a house in a small town in New Mexico; but I'm not abandoning the computer field, and I would have been moving out of New York City even if Y2K hadn't come along" He suggests, however, that Y2K was a major factor in that decision:

> I've often joked that I expect New York to resemble Beirut if even a subset of the Y2K infrastructure problems actually materialize -- but it's really not a joke. It's likely to be fairly cold on New Year's weekend, and a combination of disruptions in utilities, telecommunications, banking, schools, hospitals, airports, unemployment checks, Social Security checks, food stamps, and/or welfare checks would be enough to make the citizens of New York (who normally only have to tolerate problems like subway strikes and embarrassingly incompetent baseball teams) extraordinarily grumpy. There's enough gunfire in the streets even in normal times, and I'm not comfortable exposing my family to the city's ill humor if Y2K turns out to be a serious problem.

Yourdon positively agonized over the social responsibilities of programmers who, convinced that a Y2K disaster is unavoidable, head for the hills:

not every Y2K programmer can realistically contemplate leaving town for a safe haven. Some have aging parents or other family members who categorically refuse to leave the urban environment in which they reside; and some have a combination of financial, emotional, physical, or other miscellaneous reasons that require them to stay where they are... the safe-haven debate ultimately comes down to a simple and basic question: what responsibility do Y2K programmers have to stay on the job *after* December 31, 1999? If we assume that every programmer works 18 hours a day, 7 days a week, at whatever Y2K job he/she should be expected to carry out between now and December 31, 1999, then do we have a right to expect continued loyalty beyond that point?

As if starvation and riots were not serious enough problems, some proclaimed that the Millennium Bug threatened the ultimate disaster: nuclear war. According to conspiracy theorist Daniel Perez, "the Russian 'doomsday system' is designed to be switched over to automatic mode by military commanders and only launch Russia's missiles if it senses that the city of Moscow has been destroyed. The system works with no human intervention and sends coded messages to the missile silos." This automatic launching system is only supposed to be switched on if an attack is believed to be immanent, but Perez, like all doomsday theorists, assumed that the worst-case scenario is always the inevitable one: "Since the Year 2000 begins at the International Date line in the Pacific and moves westward across Siberia and Russia, the first part of the Russian nuclear arsenal that would be affected will be the missile silos in Siberia and eastern Russia." Supposedly, computers at missile sites in Siberia, believing that Moscow has been destroyed because they cannot communicate with areas where the date is still 1999, will switch over to the automatic system and begin firing nuclear missiles at the U.S.

But wait, there's more: "On August 22nd, 1999, the GPS [Global Positioning Satellite] system is scheduled to break down," according to Perez, a statement that the U.S. Defense Department strongly disputed. Actually, what happened on that date was that the clocks used by the GPS satellites rolled over to zero, similar to a vehicle odometer that rolls over at 100,000 miles. I remember turning on my old battery-powered GPS receiver at that time to see what would happen. It was confused for a few minutes, but it recovered and then displayed the correct date and time.

"If that system isn't up on December 31st 1999, the United States would be utterly defenseless and unable to launch a retaliatory strike."

Not surprisingly, Y2K survivalist advertising became a mainstay on the *Coast to Coast AM* radio show, paying many of the bills for its nonstop conspiracy-and-UFO chatter. Indeed, many Y2K doomsday websites linked in to "www.artbell.com".

Of course, Y2K came and went with only a few scattered computer problems reported. As a disaster, it was a huge bust.

The "planetary alignment" of May 5, 2000, was next threatened as a trigger for all manner of earthly havoc. But like the "Jupiter Effect" alignment of 1982, nothing happened.

## 2003: The Zetas and "Planet X"

The "Zetas" are a supposed alien race whose messages are allegedly channeled by one Nancy Lieder of Wisconsin. (Usually she just goes by her first name). She began posting her messages from the Zetas on the internet in 1995, and soon acquired a considerable following among New Age, UFO, and Conspiracy believers. When the giant comet Hale-Bopp was discovered in 1995, Nancy proclaimed that it didn't exist - it was a disinformation campaign to distract people from the coming doom of Planet X (also known as Nibiru).

According to Nancy's website,

> The inbound Planet X was sighted at the coordinates given by the Zetas in early 2001, imaged in infrared twice in January, 2002, tracked by CCD images in late 2002 to early 2003, and thereafter photographed by amateur astronomers around the world. That a middle aged woman with a high school degree, who does not even know which end of a telescope to look into, could pinpoint the RA and Dec of the brown dwarf, Planet X, is astonishing, and speaks to the validity of ZetaTalk.

This is, of course, pure nonsense. A few photographs were produced purporting to show Planet X, but there was so little information provided that it was impossible to say what the photos really showed. Nancy boasted that people had a 68 percent success rate for seeing Planet X during April of 2003, but only for "those educated, who had done their homework and followed the imaging session, noted our words as to what to look for, and oriented themselves in the sky." Apparently other people didn't see anything. She further informs us that MJ-12, the supposed ultra-secret crashed UFO panel, has "committed suicide to prevent itself from being misused." More than ten years after its alleged "discovery," no reputable astronomer, amateur or professional, has ever confirmed any sighting of the supposed Planet X.

According to the Zetas, Planet X was approaching earth in early 2003, to cause a "pole shift" and massive destruction on or about May 15. "The 12th Planet will be visible to the common man some 7 weeks prior to the shift, without the use of telescopes," says Nancy, meaning that it should have become visible to the naked eye around the end of March:

> Planet X will have a distinct red appearance, with a roiling tail full of moons, that are more concerned with the dance between them than any affect the solar wind might have upon them. Thus, they swirl, and look like a dragon approaching, not a straight line tail at all. Nevertheless, we anticipate NASA will explain the Planet X complex as any number of things, or rather their lackeys, who will natter the word on every Internet or media source that allows their nattering - asteroid bunch, passing comet, unusual comet, Mars closest pass in many eons, or whatever.

In other words, by May of 2003 "Planet X" would be huge and impossible to miss, but the government will attempt to convince you that you are actually seeing something else. Nancy noted on her website that in December of 2002, Art Bell interviewed Sylvia Browne, "the well-known psychic and author. And she disclosed she is no longer able to see the future beyond the summer of 2003." Bell also interviewed Ed Dames, who had been a "remote viewer" employed by the Pentagon. Dames said that "he and his remote viewing team are no longer able to remote view the future beyond next summer." From this, Nancy deduced that there will *be no future* beyond the summer of 2003. (But somehow we're still here.)

During June of 2003, Nancy posted numerous accounts of supposedly bizarre phenomena such as unexplained booms, sunspots, a ring around the Moon, and even the Sun "rising and setting in the wrong place." These were labeled as "signs of the times," and were attributed to the proximity of the dangerous Planet X. Nancy insisted that Earth's rotational stopping and flipping was indeed still going to occur, but she refused to specify the hour or date. Her followers were carefully noting the times of sunrise and sunset, and the position of the Sun going down, to see if Earth's expected careening might have already begun. But when these portents failed to

*Nancy Lieder channels messages of doom from the "Zetas"*

occur, most of her followers turned their attention elsewhere, to different weirdness.

A year later the Web site *www.zetatalk.com* talked a lot about the disasters expected in 2003 but said nothing about there being a year 2004; apparently their claim was that a disastrous planetary encounter did occur in 2003, but was covered up by NASA. By 2012 Nancy was Tweeting Planet X sightings to her followers (@NancyLieder1). "NASA and JPL still deny Nibiru has arrived, but their SOHO images show otherwise!" Also: "Use the insert from an old floppy disc as filter and see the Planet X complex!"

## 2004: Toutatis Threatens Totally

The End of the World came again in 2004. This time it was the asteroid 4179 Toutatis, which, according to no less an authority than the celebrated Swiss UFO contactee and "prophet" Billy Meier, was in danger of slamming into Earth on September 29, 2004. According to NASA, Toutatis would make a close approach to Earth on that date, passing within approximately one million miles, which is nothing on a cosmic scale, but still four times more distant than the moon.

But Meier is in contact with the space people from the Pleiades (he calls them "Plejarans"; I suppose that "Pleiadeans" sounds too unwieldy). And according to Michael Horn, who is the officially authorized media representative of Meier in the United States, the Plejarans have warned that when Toutatis is closest to Earth, it may suddenly veer off course and head straight toward us. Horn says that it will require a "pre-emptive nuclear strike" to keep this multi-kilometer-sized asteroid from slamming straight into Earth. Before you dismiss this prophecy as the ravings of a demented man, be forewarned that Meier claims a long list of successful predictions (no doubt selectively culled from a much longer list of unsuccessful ones—Meier reportedly has written thousands of pages of predictions). Horn says that Meier predicted back in 1987 that Islamic fanatics would destroy the World Trade Center in New York. If true, this information would have been of great interest to the then-ongoing Congressional investigation into the September 11 attacks: What did Meier know, and when did he know it?

Others also started to sound alarm bells. The Website of the conspiracy-oriented radio host Jeff Rense carried similar warnings. There was even a claim of a secret government missile program to save us from disaster on September 29. And the biblical prophecy expert Arnie Stanton suggested that the encounter with Toutatis meant that the Second Coming would be just a few months away. Stanton also warned that a larger, as-yet undiscovered asteroid will definitely smack into Earth sometime in 2006.

Of course, since it is not possible for a celestial object to "veer off course" unless some external force is applied to it, Toutatis behaved exactly like NASA said it would.

## 2008: Electric Asteroid Zaps Earth

The public rightfully paid little attention to the near-earth flyby of asteroid 2007 TU24, a piece of rock approximately 150 meters in diameter that flew past earth on Jan. 29, 2008 at about 1.4 times the lunar distance. There was no danger of any collision, just a rare chance to observe a fast-moving asteroid in a modest-sized telescope. But for those who read and posted to the website *www.tu24.org*, it was a matter of grave concern. As a group on the fringes of the "electric universe" movement, they were claiming that because asteroids allegedly have a powerful electric charge, the close approach of 2007 TU24 was going to zap earth with all kinds of nasty effects. (How an asteroid or other celestial object might acquire a net electric charge is never given a convincing explanation, other than the vague claim that it will somehow pick up "negative ions.") According to a video posted to **YouTube** that drew over 200,000 hits, **TU24** "could be a negatively-charged asteroid," and thus might cause "plasma discharge interference" with Earth, resulting in "earthquakes, deadly storms, and massive eruptions" across our planet. Despite several elementary-level science errors in the video (such as confusing "diameter" with "mass"), many people became alarmed, while others at *tu24.org* gleefully set out to record the expected electromagnetic effects of the asteroid's passage. "Severe weather is being reported worldwide since TU24's entry into the magnetosphere. From extreme conditions in China to Canada and the States to extreme cold in the Middle East and snow in Jerusalem. ALL SINCE THE 'DATES OF CONCERN' TIMETABLE ON THE RIGHT. Coincidence?" Well, snow and cold are not exactly unexpected during winter.

But **Michael Goodspeed**, writing on the ***Thunderbolts*** website decries the "slick" presentation of *tu24.org*, taking pains to note that this group has nothing whatever to do with the responsible proponents of the Electric Universe theory. As Steve Smith, another Thunderbolter, explained about 2007 TU24, "the object is far too small, plasma in space is far too diffuse, and its encounter with Earth was at too great a distance for any such phenomena to appear." In other words, TU24 will have no observable effect on us not because it isn't electric, but because its spark is too small.

According to the Thunderbolts group, astronomers should do away with not only the newest theoretical concepts such as dark matter and dark energy, but with neutron stars, Black Holes, and indeed the Big Bang itself. In *Thunderbolts of the Gods* David Talbott and Wallace Thornhill write,

> We contend that humans once saw planets suspended as huge spheres in the heavens. Immersed in charged particles of dense plasma, celestial bodies "spoke" electrically and plasma discharge

produced heaven-spanning formations above the terrestrial witnesses…. Around the world, our ancestors remembered these discharge configurations in apocalyptic terms. They called them the "Thunderbolts of the Gods."

If this sounds vaguely familiar, it might be because the "electric universe" is essentially Velikovski's universe, without using the V-word. (Exactly as old-fashioned "Creationism" has been repackaged and given new life as modern-sounding "Intelligent Design.") In 1950, the eccentric psychiatrist Immanuel Velikovsky (1895-1979) published *Worlds in Collision*, claiming that a few thousand years ago Venus went careening around our solar system like an energetic pinball, causing all kinds of "miracles" that were recorded in the Bible. (Velikovski provides "explanations" for Old Testament miracles only, since he was Jewish, not Christian.) Velikovsky knew nothing about astronomy or physics, but acquired a following that was as stubborn as they were ignorant and loyal. As is usually the case with such grandiose theorizing about ancient history, the really interesting stuff occurred well before any written records of it were made, and the phenomenon of whispering, careening planets seems determined to never occur again in this age where practically everything gets recorded.

By way of proof, the Electric Universe authors offer up numerous Native American and other ancient rock carvings of a squatting man that are claimed to greatly resemble a "plasma instability." Apparently these ancient peoples witnessed "an episode of high-energy plasma incursion into Earth's atmosphere" and made drawings of it, which unenlightened anthropologists mistook to be squatting human figures. Other familiar drawings from antiquity, including the famous Athenian Owl, are plasma discharges, as well.

Of course there are no regions of "positive charge" and "negative charge" in our solar system. What is never explained is why the "positively-charged" sun doesn't suck in all of the "negatively-charged" comets directly into it, faster than a speeding bullet.

Perhaps the most astonishing "discovery" of the Electric Universe is that

> stars are not thermonuclear engines! This is obvious when the Sun is looked at from an electrical discharge perspective. The galactic

currents that create the stars persist to power them. Stars behave as electrodes in a galactic glow discharge. Bright stars like our Sun are great concentrated balls of lightning! The matter inside stars becomes positively charged as electrons drift toward the surface. The resulting internal electrostatic forces prevent stars from collapsing gravitationally and occasionally cause them to "give birth" by electrical fissioning to form companion stars and gas giant planets.

Thus the sun and other stars are supposedly not nuclear furnaces, but rather electric heaters, warmed by supposed massive electric currents circulating across the Milky Way, their energy apparently coming from nowhere.

These same authors in *The Electric Universe* dismiss the cosmic microwave background radiation, cited by astronomers as clinching the proof of the Big Bang, as "simply the 'hum' of the galactic power lines in the vicinity of our solar system." Astronomers are likewise deluded by current theories to assume that the cratering we see on all solid bodies in our solar system are caused by impacts of smaller solid bodies (our earth being somewhat unique in that its surface keeps getting "recycled", although a few craters can still be found). Properly understood, these craters are caused by "electrical arc scarring." Talbott and Thornhill explain that "minutes or hours of electrical scarring can produce a surface like that of the Moon, which is later interpreted in ad hoc fashion to be billions of years old."

Talbott and Thornhill write (*Thunderbolts of the Gods,* Chapter 1),

> For centuries astronomers assumed that gravity is the only force that can give birth to stars and planets or can direct the motions of celestial bodies. They assumed that all bodies in the universe are electrically neutral, comprised of equal numbers of negative and positive particles. With this assumption astronomers were able to ignore the extremely powerful electric force. It was a fatal mistake.

Actually, it's no mistake to assume that two bodies are electrically neutral when gravitation alone, the weakest fundamental force, fully explains their motions, as well as that of every other visible object in our galaxy. Even a small deviation from electrical neutrality would result in massive forces of Coulomb repulsion or attraction far overpowering the gravitational force, and this simply is not occurring. At any given time, there are always a few comets in the inner solar system. If these bodies had a negative charge as is claimed, Coulomb's Law dictates that they would repel each other across great distances, far overpowering the effects of the sun's gravity.

Physicists tell us that the electrical force is $2.27 \times 10^{39}$ times stronger than the force of gravity. This is an enormous difference, the electrical elephant vastly overpowering the gravitational flea. Yet the motions of every celestial body we have ever seen are explained by gravitation alone, dutifully obeying Newton while ignoring Faraday and Coulomb. How is it possible that the much weaker force operates to the exclusion of the overpoweringly stronger one? Proponents of the Electric Universe respond that interplanetary space is actually a plasma, and plasmas act to "hide" charged particles from each other. True enough. However, the density of the ionized interplanetary medium is so thin, and the supposed electrical charges of asteroids and comets so large, that this would not matter. The answer is simple: celestial bodies are electrically neutral with respect to each other, to a uniformity far greater than one part in $10^{39}$. And the "Electric Universe" is codswollop, with or without Gods hurling Thunderbolts.

## Judgment Day: May 21, 2011

Radio evangelist Harold Camping of *Family Radio* whipped his followers up into frenzy by his confident prediction that Judgment Day, and the Rapture, would begin on May 21, 2011. This created big excitement, and eventually became a huge news story. The fact that Camping had issued an equally confident prediction of Armageddon for September 1994 did not seem to count against him

Camping, whose Family Radio broadcasts out of Oakland, California were carried on 55 radio stations in the U.S. and translated into 48 foreign languages, said he has scrutinized the Bible for almost 70 years and developed a mathematical system to interpret prophecies hidden within it. He noticed that particular numbers appeared in the Bible at the same time particular themes are discussed. This led him to conclude that certain themes are represented by a certain number. For example, 5 represents "atonement," 10 is "completeness," and 17 represents "Heaven." His predictive formula involves taking the date of Biblical events, and adding to them numbers derived from these themes. (Just in case you were wondering, the Creation occurred in the year 11,013 BCE and the Flood in 4990 BCE.) He found that the date of Jesus' crucifixion, to which is added (Atonement x Completeness x Heaven), squared and multiplied by the number of days in a solar year, gives us the year 2011! A bit more tweaking, and we get May 21 of that year. Truly, I am speechless.

*Followers of Harold Camping purchased huge billboards in many cities to warn of the impending "Judgment Day."*

A few days after the world didn't end, Camping explained to his followers that the Second Coming of Christ did in fact occur on May 21, but it was a "spiritual coming." (This excuse has been used before when a predicted doomsday failed to occur.) This began, he said, a five-month period leading up to the final judgment and destruction on October 21, 2011. Camping's original prediction of Judgment and Rapture on May 21 allowed the un-raptured to remain on earth until October 21, the date of their final destruction, so Camping is claiming that his original prediction was still on track. "It wont be spiritual on October 21st," said Camping. "The world is going to be destroyed all together, but it will be very quick."

# Comet Elenin Menaces Earth, 2011

The world was scheduled to end again in September 2011 when Comet Elenin (C/2010 X1) came to perihelion (its closest approach to the sun). This information (more accurately, misinformation) came to us from one Laura Knight Jadczyk, who channels messages from the Cassiopeans. Superficially this looks very similar to Nancy Lieder, who channels doomsday messages about Planet X from the Zetas. However, Ms. Jadczyk emphasized in April, 2011 that Comet Elenin is not Planet X, and it is not going to collide with earth. The Cassiopeans explain that they are not strictly speaking extraterrestrials, but rather in some way our "future selves."

Elenin was a faint comet, then visible only with difficulty in large amateur telescopes but was predicted to become visible in binoculars by the fall. Why was it of so much interest? For some conspiracy-oriented websites, it's primarily that the media had been strangely silent about the comet. For Ms. Jadczyk, it's because the Cassiopeans told her that the outburst of 596 Scheila, an object believed to be an asteroid that began outgassing and revealed itself to be a comet, was related to Comet Elenin in some unspecified way. She also ties this in with James McCanney's Plasma Theory of comets, and similar "electric universe" theories about comets being "electric." So when Comet Elenin reached the inner solar system, it was supposed to zap all of the planets with huge lightning bolts. Ms. Jadczyk stated that the greatest danger to earth would be on Sept. 27 and Nov. 23 of 2011, although the comet was never expected to get closer than about 21 million miles from the earth. This is indeed closer than the Earth ever comes to Venus or Mars, but still about 80 times the distance to the moon. And it was going to pass closer to the sun than Venus.

Those who were expecting mayhem and destruction from puny Comet Elenin must have been terribly disappointed when *Sky and Telescope* magazine reported on its website August 30, 2011, "**Comet Elenin Self Destructs**." This was not exactly surprising, as the same thing has happened to many small comets before as they pass close to the sun. As astronomy writer Kelly Beatty reported,

> Within the past week the comet's brightness has declined by 50%, dropping a half magnitude between August 19th and 20th... images show Comet Elenin's bright core becoming elongated and diffuse —

the telltale signs that its icy nucleus has either broken in two or disintegrated altogether. One veteran comet-watcher who's not surprised is John Bortle. Four months ago, based on Elenin's performance to that point, he cautioned, 'The comet may be intrinsically a bit too faint to even survive perihelion passage.'

That is exactly what happened. Those who would not believe the descriptions of Comet Elenin as a feeble, puny visitor to the inner solar system must have been astonished when the comet's icy core simply melted as it had its close encounter with the sun.

## 2012: Mayan Mayhem

The Granddaddy of all world-endings occurred in 2012, when, according to a number of supposedly reliable experts, the Mayan Calendar was supposed to simply "run out," apparently making it impossible for time to continue. Or so some people said. Other people, however, contended back in 1987 that the Mayan Calendar was ending then. As I wrote in my Psychic Vibrations column (*Skeptical Inquirer*, Winter, 1987-88, also *Psychic Vibrations* book, p. 213), "According to some astrologers, the ancient Mayan calendar, after allegedly counting more than 6,000 years (meaning the Mayans must have started it a few years before Creation Week, if Bishop Ussher's chronology is correct), came to an end on August 16, 1987."

Some people claimed that the 2012 Doomsday was supposed to result from the arrival of Nibiru, or "Planet X." But that is a made-up object. We recall that it was supposed to destroy the world in 2003, but failed to show up.

Lawyer Peter Gersten has been a well-known figure in UFOlogy circles at least since the 1970s. He filed suit against the CIA in U.S. District Court in 1977 for release of documents concerning UFOs. This resulted in over 900 pages of documents being released, although there was little in them that wasn't already known. There were 57 pages that were held back due to "national security" concerns. UFOlogists made a big stink about this, claiming it's proof of a government cover-up. However, the files were held back because they might allow other nations to gain information on U.S. capabilities in electronic and signals intelligence. For example, there was one document where the CIA listened in on a Cuban pilot discussing a

UFO sighting. The problem wasn't the UFO sighting, it was that we didn't want the Cubans to know we'd been listening in to their pilots' conversations.

Now retired to Sedona, Arizona, the center of a supposed 'New Age energy vortex', Gersten said that he planned to take a "leap of faith" from Bell Rock, a well-known vortexy place in Sedona, at the moment of the solstice. Gersten wrote on his website,

> On the Winter Solstice of 2012 at exactly 11:11 UT a cosmic portal will open in Sedona Arizona and a leap of faith - from the top of Bell Rock - will propel me through its opening.... My two principal beliefs are: 1) that our reality is an intelligently designed cosmic holographic program and 2) that on the Winter Solstice of 2012 at exactly 11:11 UT – a Trans-Dimensional Event (T-DE) will occur... an article on my web site discussing the Mayan calendar end-date states that: There is no reason not to take a leap of faith imagining what may be in store (10). So by now you must have an idea of what I plan to do on the Winter Solstice of 2012 at exactly 11:11 while on the top of Bell Rock – a place only a few miles from my home.... Most of you will think that I am delusional and that my insane act will certainly result in my death. Death is inevitable - at least nowadays - and 100 years from now it won't matter whether I died in 2012 or 2013 or even 2020. But I believe that some type of cosmic portal will be opening at that time and place and that an opportunity will present itself. I fully expect that it will either lead to the next level of this cosmic program; freedom from an imprisoning time-loop; a magical Martian-like bubble; or something equally as exotic. In March 2012 I will reach 70 years of age and nine months later we arrive at the cosmic coordinate. I think it will then be time for me to move on - in one form or another. I'd like to see what else our Cosmic Computer has to offer.

When I first read about this, I could not believe that it was true. Perhaps somebody with a grudge against Gersten made this up? Yet there was a video on YouTube where we see and hear an **interview** with **Peter Gersten** telling, in his own words, about his plan to (hopefully) leap into the portal that he believes will be opening up. He explains that, even if he dies in this leap, it is probably good Karma to die at the time of such special cosmic energy. He thinks that some "trans-dimensional event" will occur, sucking him into it, where he will perhaps meet the late members of the Heavens Gate cult, still wearing their sneakers, as well as Jimmy Hoffa and Judge Crater.

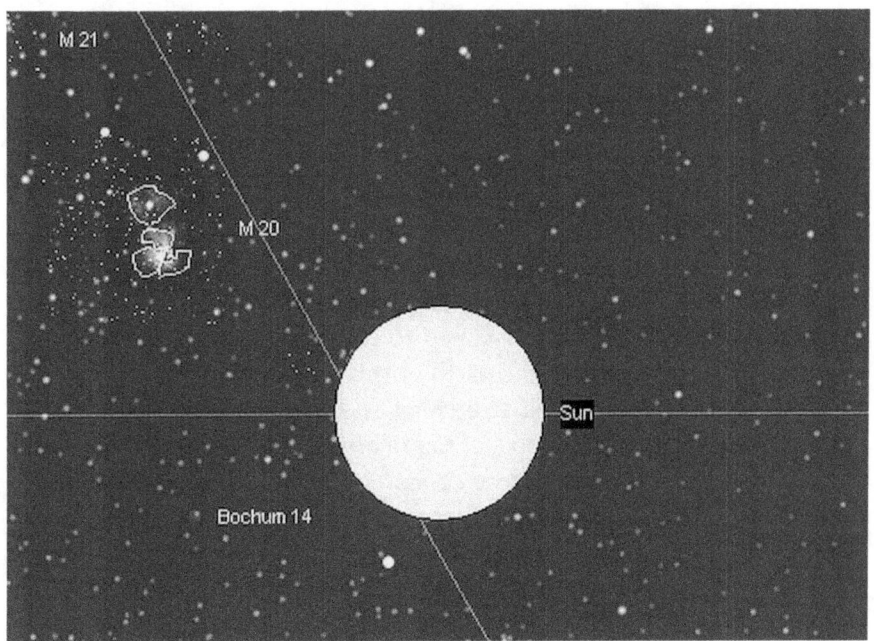

*The Sun at 2012 alignment time: close, but no cigar.*

What Gersten and so many others did not seem to realize is that from an astronomical standpoint, there was nothing special or even a little bit unusual going on in December of 2012. Nada. There would be a Transit of Venus across the Sun on the afternoon of June 5, 2012 (the morning of June 6 in Asia and Australia), which is a really special and rare celestial alignment (the next one won't occur until 2117), but clearly that is not what the hoopla was about. No "alignments" were going to happen around the 2012 winter solstice that do not also happen every other December. Supposed "Bible Codes" and "ancient calendars" simply don't cut it. No matter how many times such nonsense was repeated, it still remained nonsense. Echo chambers do not contribute to truth.

**A Galactic Alignment?** Peter Gersten posted a comment on my BadUFOs Blog disagreeing with that statement, containing a link to one of his web pages, which in turn has a link to "What is the Galactic Alignment" by John

Major Jenkins. And Jenkins had a point. There was an alignment - however, it had no significance, and it actually occurred in 1998.

To understand what this is about, you will need to visualize the celestial equator, the ecliptic, and the plane of our Milky Way galaxy. It's hard to visualize this, because hardly anyone studies spherical geometry any more (navigators of old knew it solidly). Let's use an earth globe to represent the "celestial sphere," which of course isn't a real object, but it looks like a sphere, and it can be modeled as one.

First visualize the equator. The equator makes a "great circle" around the earth, meaning that its center is the center of the sphere. (Lines of longitude are Great Circles, but parallels of latitude are not.) Then visualize the ecliptic, which is inclined to the equator at an angle of 23.5 degrees. This is the plane of the earth's orbit projected into the sky, and it also a Great Circle. Imagine it stretching on the globe from the Tropic of Cancer in Morocco to the Tropic of Capricorn in Australia. But there is yet a third Great Circle we need to be concerned with: the plane of the Milky Way, which is inclined about 60 degrees to the ecliptic.

Since all Great Circles intersect at two points, it's not a question of if they cross, but where. And the Grand Cosmic Alignment of 2012 is illustrated by this above illustration I made using Skychart (Cartes du Ciel, a free Open Source program that I highly recommend). It shows the Sun at the time of the December Solstice on Dec. 21, 2012 at 11:11:11 UT. The ecliptic is the line that's level, on which the Sun appears to be moving from right to left. The inclined line shows the plane of the Milky Way. There is your alignment.

"But wait," you say, "they're not really aligned." True enough. The actual "alignment," such as it is, occurred in 1998, when the center of the sun aligned as closely as possible with the plane of the galaxy at the time of the solstice. In fact, since the Sun has an apparent diameter of a half-degree, as seen from Earth, "alignments" such as these began in 1980, and will continue until 2016. So there you have your "Grand Alignment," which actually occurred in 1998 and was meaningless even then. Why this supposedly had any connection with the 2012 solstice was anyone's guess.

# Cosmic Doomsdays

As for the talk about the Sun "aligning" with the center of the galaxy, well, it never happens. The galaxy's center does not lie on the ecliptic, so the sun never reaches it, although each December Solstice the Sun passes only about six and a half degrees from it. Wow!

So if I were Peter Gersten, I would have waited for a much better "alignment" than this before making plans to leap off a cliff.

But when I posted this information to my Blog, Gersten disagreed once again, and pointed me to the Really Good stuff: **Thomas Razzeto**'s claims about the supposed alignment of the solstice point with the Great Rift of the Milky Way, also (allegedly) known as the **Maya Birth Canal**. This, says Razetto, constitutes "The Sacred Triple Rebirth of the Sun."

The "Great Rift" or "Dark Rift" of the Milky Way is simply big old cloud of dust and gas along the Galactic plane that obscures the stars behind it. Many galaxies have this, not just ours. Since we have seen that the Sun was not aligned with the Galactic Equator at the time of the 2012 winter solstice (although it would make no difference even if it were), if you look around hard enough, you'll always find something that the Sun is aligned with. In this case, it's the Milky Way's Great Rift. Throw a few planets into the mix, and you can fashion up a "Sacred Tree." Razzeto apparently doesn't care that Pluto has now been (quite properly) downgraded to the rank of the minor planets. Is he going to include Ceres, Vesta, Sedna, and Chiron in his "alignment" as well?

I think any rational person can see that this "Sacred Triple Rebirth" of the Sun and the "Sacred Tree" is a load of horse manure, especially when it comes from a website calling itself "infinitely mystical." I could make up a similar "cosmic alignment" story for practically any equinox or solstice, throw in a few planets, and invent some high-sounding reason why it signals the beginning of a Cosmic New Age. Most people don't realize that the inner planets Mercury and Venus spend a lot of time (from the earth's perspective) hanging out in the vicinity of the Sun, and since Mars is on the opposite side of the Sun from us at that time, it appears to move more slowly than at other times, and thus, yes, also seems to loiter for a long time in the vicinity of the Sun. To find any significance whatsoever to having Mercury, Venus, and Mars in the vicinity of the Sun at some random time requires one to be ignorant of planetary orbits. But forget all these facts: this is astrology, plain and simple, and astrology is ancient superstition, nothing more. It doesn't matter where this planet is, or that one. It's all humbug. It's just the same old "when the Moon is in the Seventh House, and Jupiter aligns with Mars." If you had gotten all worked up over a "Sacred Tree" superimposed over the "Maya Birth Canal," eventually you'd feel rather silly about it. Or at least I would.

Many people were concerned about Gersten, hoping he didn't throw away the rest of his life because of babblings about "ancient calendars" and "trans-dimensional effects." On January 26, 2012 he posted, "An Uncontrolled Growth of Abnormal Cells - Today I was diagnosed with an unusual form of cancer. An ironic start to 2012 for me don't you think? I assume my programming is ensuring that I complete my leap of faith. Bring it on! Stay tuned! My story is getting very interesting." His only related posting after that was on April 6: "My 70th Birthday Present: The Mark of the Dolphin." He explains how he went swimming with the dolphins at the Atlantis Resort in the Bahamas, and one of them bit him on the right hand. "It was bleeding and there were three 1/2-inch deep scratches with a smaller one next to them." He later realized that the dolphin was trying to write "1111" (the time of the solstice) on his hand. He suggests, "Could it be that I will need the "dolphin stamp of approval" to get through the portal?"

On November 17, 2012 I contacted Gersten by email, asking him if he had perhaps changed his plans. He said that he would go up Bell Rock at the appointed time, but would not leap unless he saw an "extraordinary event" occur - something "supernatural." I was relieved to hear that, because I did not expect any supernatural events to occur. Gersten told me that he would be atop Bell Rock by 11:00, and stay at least until midnight. I reminded him that the solstice would be at 11:11 UT, which is 4:11 AM in Arizona. He replied that the important part was not the time of the solstice, but the "symbolism" of 11:11.

On December 20, 2012 – the day before the solstice - Gersten posted to his Facebook page: "And for everyone who still thinks I will be jumping off the top of Bell Rock, please read my lips: I AM NOT! I just rented a magical place by Red Rock Crossing for 2013." He obviously couldn't move into a new house if he jumped off the rock! He wrote that he had been requested by some members of the county Sheriff's office to meet them at the bottom of Bell Rock that morning, to hike up the rock and discuss his plans. Obviously they were concerned that he might be suicidal, and I have no doubt that they would have placed him on something like an involuntary 72-hour mental health confinement if they thought he was really planning to jump. Their meeting was, by all accounts, positive and friendly. Later that day, the local newspaper *The Daily Courier* reported

> Peter Gersten says he's taking a "Leap of Faith" on top of Bell Rock near Sedona today, but that doesn't mean he's literally jumping off the prominent 547-foot-tall red butte in northeast Yavapai County.
>
> "I don't think it's going to involve risking my life," he said. "The Leap of Faith was bringing 120 people to the top of Bell Rock without knowing why." He said he felt like he needed to bring at least 111 people to the top of Bell Rock over the last several months to prepare for today.

One might legitimately call that 'moving the goalpost,' but if it's done for the purpose of preserving a human life I will not complain. As of this writing, Gersten is still alive and well and posting to the internet.

## "Blood Moons" 2015

A perfectly ordinary and predictable series of total lunar eclipses in 2014-2015 was hyped by some prognosticators as "Blood Moons." Some saw it as the "signs in the heavens" of religious prophecy. A preposterous documentary movie released in 2015, *Blood Moons*, tied it all in with religious prophecy and supposed miracles. The final "Blood Moon" in that series occurred on September 27, 2015, but nothing happened.

In early October, 2015 Chris McCann of the E-Bible Fellowship of Pennsylvania, predicted that the world would end on October 7, based on the teachings of the late doomsayer Harold Camping. McCann reckoned that Camping had the right idea, but his calculations were just a little off. "According to what the Bible is presenting it does appear that 7 October will be the day that God has spoken of: in which, the world will pass away.. It'll be gone forever. Annihilated." By fire, even. Oops!

## 2016: 'Planet X' Returns

According to a number of conspiracy theorists, Planet X was next scheduled to menace the earth in March, 2016. According to the website http://www.nibiruupdate.com/, "From our research, we think Nibiru will appear in December 2015 and pass over at the end of April 2016." This presumably is very bad news. But Alfred Lambremont Webre of *Exopolitics.com* suggests "Planet X/Nibiru may be catalyst for positive change, 2016-23."

> The stereotypical catastrophobia meme surrounding a flyby of Planet X/Nibiru, a reported Brown Dwarf star twin to our sun, is that of the "Destroyer", of a global coastal event accompanied by tidal waves destroying all coastal cities, massive solar storms triggering a pole shift, and a mega-die-off of the human race on the surface of the planet, while the monarchical, financial, political, and military-intelligence complex survives in deep underground military bases (DUMBS). Yet what if an upcoming flyby of Planet X/Nibiru that newly released data is suggesting will start in 2016 is in fact a catalyst for positive change, rather than a "Destroyer"?

## 2017: Reptilian Invasion

Having survived all these Apocalypses, the world is again scheduled to end in 2017. "**Invasion 2017**" was trumpeted by **Jerry Pippin**'s popular UFO-related internet radio show. It promised to spill the beans on "a huge armada of UFOs coming to Earth in 2017," with "complete desecration of the ETs including Reptilians, The Tall Whites and the Conformers" (I believe Pippin meant "description," not "desecration," but that's what it says.) "In addition to the Pickering Brothers we will present testimony by Charles Hall of his encounters with the Tall Whites while serving in the US Air Force at Nellis Air Force Base and other testimony by those who claim they know about the secret UN meetings with ETs and other events around the world." 2017 - Are you ready?

# 10. UFO SKEPTICS ARE FROM MARS, UFO PROPONENTS ARE FROM VENUS

After one has been in the "UFO business" for a while, one realizes that, with a few rare exceptions, nobody on one side ever converts somebody on the other side. Despite all of the "serious" arguments back in forth, no matter how much dialogue may pass between them, skeptics remain skeptics and the proponents remain proponents. Is it because of deeply held religious-like beliefs, beliefs that one is unwilling to question? In some cases, proponents do hold to UFO beliefs as a surrogate religion, beliefs that play much the same role in their lives as Bible belief does to the pious. But this is by no means the complete story. Many people of obviously high intelligence accept things that seem to violate what we know about physics, yet defend those beliefs with sophisticated argument

I got to know J. Allen Hynek when I was a student at Northwestern. I took several classes from him, and I was able to participate in a number of UFO-related discussions with him. He would sometimes invite his students into his home, to discuss whatever they might like. I went there several times. Hynek was a strong UFO proponent at that time, yet obviously highly intelligent, and well-able to defend his beliefs. Over the years I have also gotten to know a number of well-known UFO proponents, including James Moseley, Bruce Maccabee, Kevin Randle, John Alexander, and Karl Pflock. All of them were highly intelligent, and quite capable of giving plausible-sounding reasons for their belief that UFOs are anomalous and real. What is it that separates sophisticated UFO proponents from skeptics, the Sheep from the Goats?

## Overconfidence in the credibility of "Reliable Witnesses"

This is far and away the greatest difference between the Sheep and the Goats. UFO proponents generally place very high credibility in the "testimony" (strictly speaking, "anecdotes") of those who claim to have had experiences involving UFOs. Generally speaking, if a person appears to be "credible" (that is, apparently stable, educated, and socially well-adjusted), then his or her "testimony" will be considered as something akin to proof, even if it describes objects or events that seem to be impossible. The fallacy being committed here is obvious to anyone who actually understands the history of paranormal claims. Going back just a few centuries, one can find "testimony" of the very highest caliber affirming the existence of witches who could perform supernatural feats, and of various religious miracles. Joseph Glanvill, ever the good empiricist, noted the solid character of the testimony affirming witchcraft (Glanvill, 1689, p. 67):

> In order to the proof that there have been, and are unlawful confederacies with evil spirits, by vertue of which the hellish accomplishes perform things above their natural power; I must premise that this being matter of Fact, is only capable of evidence of authority and sense; and by both of these, the being of Witches and Diabolical contracts, is most abundantly confirm'd. All Histories are full of the exploits of these Instruments of darkness; and the testimony of all ages, not only of the rude and barbarous, but of the most civiliz'd and polish'd World, bring tidings of their strange performances.

Even today we have accounts from apparently highly credible persons who claim to have witnessed extraterrestrials, angelic beings, ghosts, Bigfoot, the Virgin Mary and other religious figures, fairies, etc. All of these are things that are believed by science to not exist, at least not in our normal, 3-dimensional "real world." Yet here we have "testimony" of those who affirm it. Examining the true scope of the conundrum, we see that if we open the door to belief in UFOs based on "reliable witnesses" alone, we will be unable to keep witches, fairies, ghosts, Bigfoot, etc. from squeezing through that crack. Some UFOlogists understand the problem and have come up with hypotheses that are even more bizarre (for example, Jacques Vallee's *Passport to Magonia* or John Keel's *ultraterrestrials*). There is a continuity of UFO reports on the spectrum of weirdness including all of the seemingly-impossible things that have been reported throughout history. This is an example of "the old lady who swallowed the fly" solution: trying to deal with something you have swallowed that is hard to digest by swallowing something that is worse still.

## Failure to Apply Occam's Razor, and to Fully Explore the Consequences of the Assumption

Very seldom does a proponent of UFOs or other weird things properly apply Occam's Razor to the situation: the search for the most parsimonious solution to a problem, one that does not require the "multiplication of entities." Or as skeptic Dr. Shawn Carlson explained Occam's Razor, "the most boring explanation that fits all the facts is the most likely to be correct." For example, when somebody reports seeing Bigfoot in the woods, to accept that account as authentic requires we postulate the existence of a new entity, a giant hominid that is especially adept at avoiding unambiguous detection, with absolutely no evidence of it in the fossil record, or in the forest. Unless we have solid physical evidence of such a creature, this is what Occam's Razor tells us not to conclude.

Typically there is also the failure to examine the consequences of trying to merge the newly-accepted entity with the rest of science. For example, if giant hominids exist in the forests of North America, how large is the breeding population? What is their food? Where are their remains? Where are their droppings? Where are their fossils? How are they related to known ancient hominids? When any of these questions are followed rigorously toward a solution, they show it is impossible for science to accept the new entity you are proposing without making radical revisions to what is already known about the biology of mammals. Is it wise to radically revise much of biology because we have received anecdotes about unknown hominids?

## Failure to Understand Important Facts in Astronomy or other Sciences

A person may be highly educated and accomplished in one academic field, but that does not mean that he or she will necessarily be well-versed in others. For example, except for those who have chosen to study astronomy either in school or in self-study, practically no one really understands the distance scale of our solar system, the galaxy, or indeed the entire universe. They reason that since our modern ships can set forth from any port on earth to arrive in any other port in a few weeks at most, some day our astronauts will be traveling back and forth to the stars in this same manner, as we see on *Star Trek*. This gives rise to the *Prime UFO Fallacy*, discussed in chapter 8: that to deny the reality of UFO sightings here and now is the same as denying the possibility of intelligent life elsewhere in the universe. Nothing, of course, could be further from the truth. The near-impossibility of interstellar travel, demonstrated in that chapter, shows why the 'terrestrial navigation' analogy for interstellar exploration is an extremely bad one. It also shows how it may be possible that the universe contains large numbers of intelligent civilizations, some able perhaps to communicate with others, but none able to bridge that all-but-impossible gulf by travel.

Also, the movies and television have accustomed us to seeing so many impossible events associated with UFOs and alien visitors that we seldom recognize them as impossible. For example, if a landed UFO simply rises from the ground and zips away, as is often claimed, we have some extremely serious problems. We have an action – the object allegedly rising – without an equal but opposite reaction (expelling mass). This violates Newton's laws of motion, and the conservation of momentum. When a "witness" relates an anecdote such as this, implying that much of what we know about physics is wrong, I must assume that the person is confused or untruthful instead of accepting their unsupported and impossible tale.

Another impossibility often not recognized concerns supposed "alien hybrids" as reported by many so-called "UFO abductees." To make a "hybrid" between two living creatures requires that the chromosomes with their DNA line up to a high degree. This requires the creatures to be closely related. Sometimes you can make a hybrid between closely related species; the dividing line between species often is not hard-and-fast. But you cannot make a cross between a horse and a horse-fly, even though their genes may match by 50%, because there are simply too many differences in the organisms. So imagine the difficulties encountered in trying to match up human and alien DNA. Assume that you have encountered an alien species that has evolved on a distant planet. They would presumably have evolved genes to perform many of the same functions as ours do, but arrived at by a completely different evolutionary path. They may, or may not, have evolved DNA. To expect to make a hybrid from these two completely unrelated species – the Grays, and us – is absurd. By comparison it would be much easier to make a hybrid between a horse and a horsetail plant, where you might find a 25% match in genetics, which is much better than zero.

## Personal Experience

Like those who believe in a personal God, psychic powers, ghosts, etc., UFO belief is often grounded in an experience so powerful that its validity cannot subjectively be denied. A person may have had a sighting of something in the sky that was so puzzling, and its impact so overwhelming, that the person is convinced that it must have been from another world. As time passes, the person thinks about this event, again and again. It ultimately ends up as a life-defining experience, a litmus test by which all other beliefs are measured. If science says that extraterrestrial UFOs do not exist, then science must be wrong, because I have seen one!

Of course, people who have had such intense personal experiences seldom understand the **unreliability** of human **eyewitness testimony**, the **malleability** **of** **memory,** **sleep-related** **hallucinations** and misperceptions, and the mind's tendency to interpret the unfamiliar in terms of something that is familiar (such as a "flying saucer").

## Psychological and/or Religious Factors

Some individuals are compelled by powerful psychological, ideological, or religious factors to reach irrational conclusions. Some people experience a powerful inbuilt, reflexive opposition to any hierarchy, even if it is a meritocracy of the intellect. Often there is a powerful envy and resentment against science; the more science is esteemed by some, the more it is opposed by others. I wrote a great deal about this in my 1988 book *Resentment Against Achievement*. Such a person is delighted by claims that science is wrong, and that the ideas of the simple folk are right – like extraterrestrial flying saucers! The idea of science being completely wrong is so satisfying that the resentful will almost automatically embrace claims affirming it.

Others have lost their beliefs in traditional religion, but still retain a longing for the mysterious and the transcendent. Belief in extraterrestrials – especially powerful and benevolent ones – fulfills this need perfectly. It gives the universe a sense of meaning that seems to be lacking in pure science. Some people seem to require the existence of an imaginary world, very different from our earth, as part of their intellectual universe.

## "Reliable Witnesses"? Take Nobody's Word for It!

In our day, we still receive reports from seemingly credible persons of things we think almost certainly do not exist: UFOs, Bigfoot, angels, and miracles of every kind. The popular press and the mass media are very fond of such claims. They make for good ratings. But they make for very bad science.

We can use cases like the Phoenix Lights as an example to illustrate the inverse relationship between the amount of evidence presented for a particular UFO case, and its apparent mysterious-ness: the more people who see, and photograph, a supposedly "unex-plained" object, the more easily explained it will be. In this case, it's a flare drop. It is those sightings that take place in distant places, or are seen only by one or a few, that will make the skeptic work harder for a solution.

*The Royal Society of London for Improving Natural Knowledge*, invar-iably shortened to *The Royal Society,* is the oldest and probably the most

The Bookplate of the Royal Society – NULLIUS IN VERBA *(Wikimedia commons)*

prestigious scientific body in the world. Founded in 1660, over the years its Fellows have included such luminaries as Robert Hooke, Robert Boyle, Isaac Newton, Humphry Davy, Charles Darwin, and practically every other

# UFO Skeptics Are from Mars, Proponents are from Venus

British scientist of any note, as well as many foreign ones. Its motto is *Nullius in verba* – literally "on the word of no one," or more colloquially, "take nobody's word for it." This motto has served it well for over 350 years, helping to sort out "eyewitness" claims of witchcraft, miracles, and dubious creatures from the real business of science – chemistry, physics, astronomy, etc.

The UFO proponent's motto, on the other hand, would seem to be something like *Omnius in Verba*, or "words are all we've got." In fact, Hynek was honest enough to admit that directly: we possess no actual UFOs themselves, he would say, only reports of them. Hynek was surprised and genuinely hurt when the scientific establishment replied to him, as it had to in order to remain true to its centuries-old foundations, "we take nobody's word for it." Give us a piece of a UFO, or some indisputably authentic, clear and detailed photos, or some instrumental data. But if all you have to offer are stories about sightings of UFOs, we are not interested. *Nullius in verba*.

Jacques Vallee wrote that, concerning UFOs, "the scientific world is as close-minded as an old pig" (Vallee 1996 p.184). Jacques, you're a very bright fellow. You should know that *Nullius in verba* has been the rule in science since the 1660s. Since that time, the scientific world has accepted many profound new ideas such as Newton's laws of motion and of gravitation, electromagnetism, Einstein's relativity, quantum mechanics, plate tectonics, the Hubble expansion and the inflationary universe (the most bizarre and hard-to-believe scientific theory I've ever heard of, but it appears to be supported by observation). These are just a few of the major paradigm changes occurring in science since *Nullius in verba* became the rule. Not bad for a close-minded old pig. Now exactly what kind of evidence do you have to offer, sir? Words?

**That** is the mistake of Hynek, Vallee, and so many others: accepting "eyewitness testimony" as a foundation for a major revision of science, in the absence of actual physical evidence. Or, succinctly, "extraordinary claims – but little or no proof."

# REFERENCES & INDEX

## References:

Bullard, Thomas E., "*UFO Abductions: The Measure of a Mystery. Volume 1: Comparative Study of Abduction Reports.*" Fund for UFO Research, 1987.

Burke, George, 1977: "UFO Investigator Refutes Betty Hill's Recent Claims." *Foster's Daily Democrat*, Dover, N.H., October 15, 1977.

Cameron, A.G.W. (editor), 1963: *Interstellar Communication*. (New York: W.A. Benjamin, Inc.)

Clark, Jerome, 1978: interview with Betty Hill. *UFO Report*, January 1978.

Clarke, Arthur C., 1964: *Profiles of the Future*, Bantam.

Clarke, David, 2015: *How UFOs Conquered the World* (London: Aurum Press, Ltd.)

Dille, Robert C. (ed), "The Collected Works of Buck Rogers in the 25th Century" (New York: Chelsea House Publishers, 1969).

Evans, Hilary and Stacy, Dennis (editors): *UFOs 1947-1997* (London: John Brown Publishing, 1997 )

Feynman, Richard P., 1998: *The Meaning of it All: Thoughts of a Citizen Scientist*. (New York: Addison Wesley.)

Fitch, C. W., 1963: "The Experience of Mr. And Mrs. Barney Hill." (*The APRO Bulletin,* March, 1963.)

Friedman, Stanton and Slate, B. Ann: "UFO Starbase Discovered", *Saga*, July, 1973.

Fowler, Raymond: "Telepathy and a UFO." *Official UFO*, January 1976, p. 14.

# Index

Loftus, Elizabeth and Ketcham, Katherine (1996): *The Myth of Repressed Memory*. (New York: St. Martin's Press)

Lorenzen, James and Coral, 1977: *Abducted! Confrontations with Beings from Outer Space*. (New York: Berkley Press).

Marden, Kathleen and Friedman, Stanton, 2007: *Captured! The Betty and Barney Hill UFO Experience*. (Franklin Lakes, NJ: New Page Books).

Pflock, Karl T.: *Roswell – Inconvenient Facts and the Will to Believe* (Prometheus Books, 2001).

Saunders, David, 1975: *Astronomy* magazine, August, 1975.

Schwartz, Berthold E.: "Talks with Betty Hill". *Flying Saucer Review* 23, nos. 2,3,4(1977).

Sheaffer, Robert 1998: *UFO Sightings* (Amherst, New York: Prometheus Books).

Sheaffer, Robert, 2011: *Psychic Vibrations* (Charleston, SC: Create Space).

Showalter, Elaine, 1997: *Hystories - Hysterical Epidemics and Modern Media.* (New York: Columbia University Press.)

Soter, Steven and Sagan, Carl: *Astronomy* magazine, August, 1975.

Vallee, Jacques 1996: *Forbidden Science - Journals 1957-1969*. (New York: Marlowe & Company).

Wysocki, David: "Reporters join UFO trackers in N.H. Woods." *Boston Herald American*, Oct. 10, 1977.

# Index

# Index

Kelly, Mark. 189.

Kendall, Judy. 137.

Keyhoe, Donald. 9, 12, 273.

Kimball, Paul. Ix, 16, 108.

King, Larry. 13, 27

Kingman, Arizona UFO crash. 110-111.

Kinross UFO. 5.

Kitei, Dr. Lynne D. 28.

Kiviat, Robert. 29.

Klass, Philip J. v, vi, 14, 20, 21, 56-57, 114-115, 119, 125-126, 139-145, 151, 165, 197, 273.

Knapp, George 205.

Koi, Isaac. ix, 59, 108,

Kottmeyer, Martin S. ix, 8, 14, 115, 126, 134-135, 160, 192, 273.

Larson, Sandra. 137.

Lator, Nab. 106.

Latvia meteorite hoax. 113.

LaViolette, Paul. 197.

Lehmberg, Alfred. 71.

Leir, Dr. Roger. 159-160.

LeMonda, Jack UFO photo. 63-64.

Lieder, Nancy. 247-249, 256.

London UFO mothership video. 77-80.

Long Island UFO crash. 112-113.

Lorenzen, Coral and James. 4, 136-137, 141, 146, 274.

Loughner, Jared. 187-189.

Lucci brothers UFO photos. 58, 76.

Maccabee, Dr. Bruce. 265.

MacDonald, David. 6

Macdonald, James D. 120-122, 124

Mack, Dr. John. 147, 149, 157-158.

Maillot, Eric. 42.

Mallove, Gene. 197.

Majestic 12. 164-165, 172, 248

Marden, Kathleen. 120-121, 133, 274.

Markowitz, Dr. William. 230-232.

Marrs, Jim. 102-103, 212-213.

Martinek, Mary. 132, 134.

Matson, Rob. 38.

Maussan, Jaime. 10, 67-68, 71, 106-107.

Mayan Calendar. 3, 257-258.

McCann, Chris. 263.

McDonald, Dr. James E. 56, 120.

McGaha, James ix, 23-24, 26.

McGonagle, Joe. 208-209.

McKinnon, Gary. 206-207.

Meier, Billy. 249-250.

Mengele, Josef. 99-101.

Menzel, Donald H. 14.

Men In Black. 1, 61, 170,

Mervine, Kitty ix, 131.

Miller, John Lester. 65-66.

Ministry of Defence. 10-11, 43, 174, 207, 209.

Mitchell, Edgar. 165, 197.

Molczan, Ted. ix, 17, 19, 106, 176.

Mogul project. 95, 192.

Montauk Project. 185-186.

Moody, Sgt. Charles I. 146.

Moody, Lance. 107

Moore, William L. 95, 101, 164.

Mori, Kentaro. 74, 83.

Mortellaro, James. 155-156.

Moseley, James. 103, 235, 265.

MUFON. 3-11, 27, 72-73, 101-102, 110-111, 135, 153, 159.

Murphy, Dr. Tom. 225-226.

Musgrave, John Brent. 90.

Mystery missiles. 44-49

NARCAP. 32.

National UFO Reporting Center. 10, 22

Nazi saucers. 6, 99-101, 118, 121, 211-215.

Nelson, Michael. 5.

Nettles, Bonnie Lu. 240-241.

New Age UFOlogy 2-3, 247, 258, 261

Newton, Isaac. 254, 268, 270-271.

Newton, Silas. 92-94, 101-102

Nibiru. 247, 249, 257, 263-4.

NICAP. 4, 12, 61, 136.

Nichols, Preston. 185.

Nietzsche, Friedrich. 16.

Night vision equipment 32-37.

Noel, Ron. 159-160.

Noory, George. 13, 29, 167.

North, Gary. 243-244.

NOSS satellites. 33, 36.

*Nullius in verba*. 270-271.

Obama, Barack. 46, 174, 178, 216-218, 220-222.

Oberg, James E. ix, 17, 160, 165, 186, 189, 197.

O'Connor, Dr. Richard. 107.

Occam's Razor. 2, 131, 160, 162,-163, 235, 267.

O'Hare Field sighting. 31-32.

# Bad UFOs

# Index

# About the Author

*(Photo by Susan Gerbic)*

Robert Sheaffer is a writer with a lifelong interest in astronomy and the question of life on other worlds. He is one of the leading skeptical investigators of UFOs, a fellow of the well-known *Committee for Skeptical Inquiry* (formerly CSICOP), and has written the Psychic Vibration column for its publication *The Skeptical Inquirer* for over thirty-five years. He is also a founding director and past Chairman of the *Bay Area Skeptics*, a local skeptics' group in the San Francisco Bay area.


Mr. Sheaffer is the author of *UFO Sightings* (Prometheus Books, 1998), *Psychic Vibrations* (2011), *The Making of the Messiah* (Prometheus Books, 1991), *Resentment Against Achievement* (Prometheus, 1988) and has appeared on many radio and TV programs. His writings and reviews have appeared in such diverse publications as *OMNI, Scientific American, Spaceflight, Astronomy, The Humanist, Free Inquiry, Reason*, and others. He is a regular columnist for *The Skeptical Inquirer*. He is a contributor to the book *Extraterrestrials - Where Are They?* (Pergamon Press, Hart and Zuckerman, editors), which *Science* magazine called "one of the most interesting and important of the decade." He has written the article on UFOs for Prometheus Book's *Encyclopedia of the Paranormal*, as well as for the *Funk and Wagnalls Encyclopedia*. He has been an invited speaker at the Smithsonian UFO Symposium in Washington, DC, at the National UFO Conferences held in New York City and in Phoenix, as well as at the First World Skeptics' Congress in Buffalo, New York.

Mr. Sheaffer lives near San Diego, California. He has worked as a data communications engineer in the Silicon Valley, and sung in professional opera productions. His website is *www.debunker.com* , and his Blog is *www.BadUFOs.com* .